Globalization, Engineering, and Creativity

Globalization, Engineering, and Creativity

John Reader

www.morganclaypool.com

ISBN: 1598291521 paperback

ISBN: 9781598291520 paperback

ISBN: 159829153X ebook

ISBN: 9781598291537 ebook

DOI: 10.2200/S00064ED1V01Y200610ETS003

A Publication in the Morgan & Claypool Publishers' series
SYNTHESIS LECTURES ON ENGINEERING, TECHNOLOGY AND SOCIETY #3

Lecture #3
Series Editor: Caroline Bailie, Queens University, Onlario, Canada

ISSNs: 1933-3633 (Print)
ISSNs: 1933-3461 (Electronic)

First Edition

10 9 8 7 6 5 4 3 2 1

Globalization, Engineering, and Creativity

John Reader

SYNTHESIS LECTURES ON ENGINEERING, TECHNOLOGY AND SOCIETY #3

MORGAN & CLAYPOOL PUBLISHERS

ABSTRACT

The text addresses the impact of globalization within engineering, particularly on working practices and prospects for creativity. It suggests that accepted norms of economic activity create enclosures and thresholds within the profession, which—as engineers increase their awareness (reflexivity)—will shape the future of engineering, and the values which underpin it. It is aimed at practicing engineers and those in training and is an introduction to the social and political context currently setting new challenges for the profession.

KEYWORDS

Creativity, Enclosures/thresholds, Engineering ethics, Globalization, Impact of technology, Labour flexibility, Liquid modernity, New Economy, Reflexivity

Contents

CHAPTER 1

Impacts of Globalization upon Engineering

1.1 INTRODUCTION

In July 2006 the UK press reported the following story. James Dyson, famous for designing the domestic appliance known worldwide as the 'Dyson' (a type of bagless vacuum cleaner), was proposing to set up a new academy of design and enterprise in despair at the way in which UK schools were neglecting budding young engineers. The report went on to say that this was just the latest example of the drastic steps that business leaders were forced to take in order to arrest the decline in skills and competitiveness. The major fear is that China and India are about to race ahead leaving economies such as the UK far behind. Figures quoted suggest that the UK produces 24,000 engineering graduates a year, compared to 300,000 in China and 450,000 in India. Questions are raised about the government's real commitment to engineering and science in its educational policies despite its claims to be placing an increased emphasis upon these areas of the curriculum.

As well as formulating a new course on business and engineering, the academy would offer courses on Mandarin. Mr Dyson himself is quoted as saying: "The school will allow young people to explore ideas, experiment and solve real world problems. We want to encourage future generations of design engineers." Support will also be forthcoming from the aerospace groups Rolls Royce and Airbus and the Formula One teams of Williams and MacLaren. However, it was also noted that the company had actually moved its manufacturing plant to Malaysia four years ago, thus cutting 800 jobs, but that it retained a team of engineers and designers at its UK headquarters. This latest scheme is in addition to the Rolls Royce two-year project to find the most original science teachers in the country and the recent concerns about the long-term future of the aerospace industry in the UK, which is probably the largest remaining manufacturing sector along with the automotive industry.

Within this brief account it is possible to identify ways in which globalization is now impacting upon the engineering profession. Yet one must be cautious in simply attributing all determining factors to one cause or set of related causes. The argument needs to progress in careful stages, aware of generalizations and open to alternative interpretations. What I aim to present in this text are ways in which globalization creates both enclosures and thresholds for the engineering profession. Enclosures are areas where people feel trapped or constrained by developments beyond their control and thresholds being the possibilities that are glimpsed of new and exciting developments, often despite or in the midst of the enclosures. I hope in this way to guard against any form of determinism, the suggestions that 'things are bound to be this way', which invariably leads to pessimism or a fatalistic attitude towards the future. Unless it is possible to believe that things might be otherwise there seems little point opening up such a discussion. Finding a realistic balance between optimism and pessimism is surely a good survival strategy in any walk of life.

1.2 ISSUES RAISED BY GLOBALIZATION

For the moment it will be useful to identify the range of issues which arise from the Dyson story. First is that of global competition and, in particular, the rise of China and India and indeed other Far Eastern economies. This is the current background to wider concerns about the impact of globalization upon the developed economies. It includes worries about the loss of manufacturing jobs and the apparent gap in skills and education between different parts of the world. Is this inevitable or can something be done about it? We will see shortly that attempts are being made to tackle the educational dimension of this but that hopes of translating this into large-scale changes in the numbers involved in actual manufacturing are less evident. Second is the role of design and innovation within the engineering profession. Even though manufacturing may inevitably be shifting to the Far Eastern economies because of lower labour costs, is it still possible that creative and innovative research and design functions can be retained in the developed nations? The hope clearly is that this will be the case. Yet this raises a further set of issues. What sort of companies or businesses will be able to survive and flourish in this context? Will these be the larger and transnational corporations operating to the rules of the global economy, themselves determined, as we shall see, by employment practices that may or may not be conducive to creative engineering? Companies that are basically profit and shareholder driven bring their own constraints and limitations for many who are concerned about more than simply earning their living and also raise wider ethical and economic questions. Perhaps alongside this will run the capacity of smaller and networked businesses to provide spaces and opportunities for greater freedom to think creatively and critically within the profession. Does big business equal enclosure and small business equal threshold, or are both in fact locked into the same system?

What are engineers for? Is it the case that, outside of the academic world anyway, the role of the engineers is to produce the ideas and designs that contribute to the process of wealth creation and to be the raw material of further profits for big business? Is there some other sort of contribution to human well-being that engineers can still offer and ways in which their skills and experience can benefit those who lose out in the global economy? What is meant by 'globalization' in any case? This will require much more detailed examination in the second chapter.

I realize that I have begged a significant question about the role of the academy as well, particularly in countries that are attempting to foster much closer relationships between university departments and business interests. The idea that there are independent and innovative engineers and scientists pursuing research for its own sake appears somewhat naïve, although one imagines that some safe enclaves of research do still exist, which would be an exception rather than a rule. But then, what has actually driven or financed most research and innovation over the last 200 years in any case? How much has always been tied into the demands of the military and the defence industry and how much that subsequently entered the commercial world derives from those sources? Has globalization changed this in any way or is it still in fact 'business as usual', but now on a larger scale?

What is the role of government in any of this? It would appear that any government's major concern is about remaining competitive in the global economy. Hence issues about the quality and quantity of the teaching of science and engineering both at school and graduate level are simply geared to internal economic measures and have no wider vision of the role of the professions let alone an understanding that these are spheres where exciting and significant breakthroughs occur that are of more than financial benefit. One might be tempted to ask where human beings as human being come into this picture, if at all. Is inspiration and inducement into engineering really going to emerge from within such an economically determined context? Why would anybody even think about becoming an engineer? Is it for the money, or in spite of the money, or perhaps for some other set of motives? Where is the evidence for this and whose interests are being served when governments insist on playing a leading role in asking the questions?

I suspect that might be enough to be going on with, but I hope I have shown that one cannot begin to address the question of how globalization might be impacting upon engineering without being drawn into a range of issues and disciplines. Hence, I will be using material from sociology, economics, politics and even philosophy during the course of this text. I will aim to provide some tools of interpretation that illuminate the context within which engineering now operates and to offer both critical perspectives and the odd glimpse of an alternative vision.

1.3 BUILDING THE BROADER PICTURE

In order to help me gain some measure of what is happening within engineering I have drawn upon a number of sources within the UK and the USA that I hope have provided a rounded view of the current situation. If the argument is that global influences are now shaping the profession, then the picture presented from any one context should progress the discussion without claiming to be definitive or determinative. It also needs to be said that it is difficult to isolate engineering, certainly at a public policy level, from the wider debate about science and technology. So for the purposes of this section, I will be talking more generally about science, engineering and technology (SET) as they are viewed within the UK at a national level.

What emerges from this exploration into the current state of affairs is a range of concerns, some of which have links to issues of globalization and some of which are more obviously domestic. It is important to register both of these dimensions if only to acknowledge that global factors are only part of the picture, albeit now perhaps the key feature or the colours in which this picture is being painted. One might ask why it is that these concerns now appear of greater importance at public policy level and the answer would appear to be that it is about becoming or remaining competitive in the global economy.

So here are the main findings of my own research presented without too much detail or further interpretation.

The public profile of SET is apparently a cause for concern, with the image of science-based subjects not being conducive to people entering the professions in adequate numbers. We will examine the numbers shortly, but what is interesting here is the finding of a Royal Society research paper published in 2006 (www.royalsoc.ac.uk/survey), which suggests the reasons for this poor public profile. A problem appears to be that those already within the academic sector are more anxious about obtaining research funding and pursuing their own careers and research than they are in communicating the excitement and potential benefits of entering this world. The message and the vision of what a career in SET might offer are being subordinated to individual ambition or departmental needs. One might argue that this is the result of the particular UK academic culture and that would certainly be true, but one could also add that this is itself the result of political decisions about academic funding regimes themselves made in the light of the diminished role of the UK within a global academic context and the supposed need to be more focussed and target driven. If the impact of this is that younger people are not being inspired or motivated to enter these fields but instead look to more highly remunerated professions, then it is hard to see how this can be significantly altered without a complete culture change. If the wider culture is based on the value of finding meaning through financial reward, then scientists themselves are going to be subject to those values and prospective recruits will make their judgements accordingly.

Public confidence in the world of SET is also not what one might expect. Although it might be the view of those from within the profession that SET provides the most secure and trustworthy basis for the advancement of knowledge and other dimensions of human progress, this is not necessarily shared by the wider populace. This might come as a shock, but once one realizes the ways in which 'experts' are now being used by government to justify particular decisions and policy directions this becomes more easily explicable. Recent domestic examples such as the Foot and Mouth outbreak of 2001, the problems created by BSE within the farming and food industry, let alone the questions about other health issues and the role of expert advisors in the decision to go to war in Iraq, all contribute to a growing suspicion that the government selects the experts who provide the answers they want the public to hear and accept (Reader, 2005). If this is the case then it is also true that other experts will give a very different view or interpretation. So who is one to trust in all of this? If there is no one correct or trustworthy answer to so many of these supposedly objective or scientific questions, and vested interests either political or economic appear to determine answers that are presented, then why should the public place a high degree of confidence in those who operate in these areas? Of course one should engage in discussions about the difference between research and its applications, but most public debate will not reach that level. Perhaps this is not about globalization directly, but it does reflect a culture in which different world views, interpretative frameworks, or what the philosophers of post-modernity term metanarratives, exist side by side and compete for support and credibility.

One reason for this lack of confidence is the failure of SET professionals to recognize that the general public is concerned about the issues of values and of morality—this is according to a UK Parliament Select Committee Report on Science and Technology in 2000. It seems to me that this is significant and throws the gauntlet down in a very clear manner. Research cannot simply be divorced from its wider context and the questions of who funds it and how it is likely to be implemented and for whose benefit. Perhaps this is to compromise the notion of pure research, but links with political and commercial interests are now too well established in the public arena to be left out of the equation. The global media coverage achieved by environmental groups has changed the context within which SET operates and created a more hostile and critical reception for the 'experts'.

1.4 PROBLEMS OF RECRUITMENT AND TRAINING

What about the actual numbers of people coming through the educational system at different levels? The report of the Engineering and Technology Board in 2005 (Engineering UK 2005) provides recent evidence. It suggests that the increase in numbers taking GCSE (16-year-old exams) is largely attributable to demographic increases. However, there has been a 24% decline in A-Level maths students since 1995. Average grades at these exams have improved, but there are some who question whether this is simply the result of a lowering of standards, so the results are contested to say the least. More significant though are the figures for higher education. The overall number of undergraduate entrants into the system over the last 10 years has grown by nearly 40%. However, the number of students accepted into physical science courses has declined from 15,000 in 1994 to just under 14,000 in 2004. Engineering and technology have also remained static with acceptances at around 16,500. Students taking biological sciences and computer science courses have grown by 120% and 159% respectively. The number of students who then transfer out to other professions having completed their degrees is very similar to other subject areas, but the worry comes because levels of remuneration often compare unfavourably to competing employment paths. According to the Office of National Statistics the average annual gross earnings of professional engineers at the end of 2002 was £33,000, but the findings of the ETB themselves show that the median chartered engineer salary in 2005 was £45,500 (up from £43,500 in 2003).

The same report concludes that although adult perceptions of SET generally are positive, there is an out-of-date view of science and engineering and that the latter in particular gets a bad press in terms of what is actually involved in the role and is often believed to be 'dirty and hands-on'. People acknowledge the importance of the work to life generally but are disinclined to see it as a profession that they would consider pursuing or to recommend to young people. When they do so, it is often because there is already a member of the family in the profession or because there is a role model within the wider networks. The question of societal role models and 'icons' or high-profile projects which capture the imagination is itself significant and one wonders whether there is a general lack of these within the current UK culture generally as, again, SET loses out to other alternatives as presented by the global media—e.g., sports and entertainment stars.

A further concern in the UK is the teaching of SET at particularly sixth form level and the impact that having teachers who do not themselves possess appropriate higher SET qualifications now has upon the quality of education. Before so many women were able to enter either the academic or professional and business worlds, their main outlet was probably to go into teaching, thus providing a pool of committed and qualified staff. Now that has shrunk for obvious reasons. There are also questions about the way that the national curriculum in the UK is too prescriptive about the ways in which science is taught at primary school level, denying teachers the opportunity to present their own particular areas of interest in ways that are more likely to enthuse and inspire their pupils.

So a picture is starting to build of an area where external factors not always directly related to the actual content of SET are having potentially damaging impacts upon the numbers entering and then remaining within the field. Domestic culture is certainly partly responsible and has led to a number of recent initiatives to bolster the image of the profession and to improve contacts between teaching at all levels and the real life of engineering and science both within academia and industry—the work of the ETB is to be noted in this context. However, if it is cultural factors that are so significant and those themselves are simply part of a wider global culture in many ways, then one wonders how much impact they are likely to have. Terms such as values, vision and inspiration continue to be mentioned but may not actually be consistent with the economic drivers that lie behind these initiatives. What exactly is the vision which might inspire people to become engineers? Is it personal gain or perhaps the opportunity to contribute something of value to others? Of course the two are not necessarily in conflict, but are the influences of a particular form of economic globalization more likely to create that conflict? We need once more to go on enhancing the picture.

1.5 CASE STUDY: TOOLS, DIES AND INDUSTRIAL MOULDS

In order to bring the impact of globalization into sharper relief I will now concentrate briefly upon the issues facing one particular branch of mechanical engineering, that of the section of the industry known as tools, dies and industrial moulds (TDM), and a report of the United States International Trade Commission in October 2002. The Commission was given the task of investigating the current competitive conditions facing producers in the USA. This included looking at the global context of the industry and examining the principal challenges and potential implications of these for business over the near term. It appeared that there were a number of dilemmas then facing the industry: a recent downturn in the U.S. economy and its slow recovery; a shrinking domestic market due to the migration of manufacturing customers to foreign locations; excess capacity due to reduced domestic demand and the impact of new technologies; customer demands for lower prices and more services; increasing foreign competition and the rising labour-related costs. Each of these clearly reflects the growing impact of global competition upon this particular sector. The changes created by this will have their effect upon working conditions for engineers and the opportunities, or otherwise, for creativity and innovation within the field.

Examining this in a little more detail will highlight central issues for the industry. In 2002 the USA had about 7000 firms operating in the sector, with more than 90% of them employing fewer than 50 people. The businesses were concentrated in areas that had long supported extensive manufacturing activity such as Michigan, Illinois, California and Wisconsin. Many of these firms have invested in up-to-date production equipment and IT in order to increase their productivity. However, adverse conditions have resulted in significant downsizing and the exit of firms from the industry (at least 200 firms had left in the previous 3 years). So a general decline in numbers employed and overall hours worked in the industry was evident.

Since much of the operation is geared to the automotive industry (nearly 50% of tooling) and the introduction of new products, and that was itself slowing down because of economic problems, there was a knock-on effect on TDM generally. The demand for shorter lead times had given an advantage to foreign competitors, particularly in Asia, who frequently operate their plants 24 hours a day. Low cost of shipment also made it easier for producers to source these products from overseas. During 1997–2001 TDM imports from China and Korea rose by 191% and 248% respectively, although from relatively low bases. The major U.S. export markets were Canada, Mexico and Germany. China was at that point the third largest die and mould manufacturer globally, still behind Japan and Germany, with a substantial number of large, foreign-invested TDM producers and with the advantage of a low-cost, well-educated labour force and a growing domestic and international customer base, Chinese wages for toolmakers being amongst the lowest in the world.

Within the EU, Germany and Portugal were the outstanding producers, but already facing the challenges of high labour costs and labour regulations. Germany has a strong tradition of craftsmanship as well as strong apprenticeship training programmes and extensive investment in research and development. Portugal's success seems to be built upon producing for specialized customers and markets.

All of this combines to place the U.S. TDM industry in an increasingly difficult competitive market. Many responding to the Commission's survey stated that their main objective in the short term was simply survival. Reduced profit margins resulting in increased cash flow problems were making it more difficult to invest in new equipment let alone the training required to keep them competitive. It was predicted that a significant number of firms would soon be leaving the business as it was estimated that the then current excess production was running at 25–30%. One industry representative forecast a 50% reduction in the numbers of firms in the U.S. TDM industry. It seemed likely that the small and often family-owned businesses would exit the sector leaving behind a number of larger firms who could more easily remain competitive because of economies of scale. Hence the general picture was one of continued decline alongside a consolidation of the more substantial companies who were better placed to survive the hostile market conditions.

This is just one section of the manufacturing side of the mechanical engineering sector but its problems illustrate how growing competition, particularly from the Far East, is directly impacting upon the industry and therefore changing the prospects and working conditions of those in the longer established industrialized nations. A further example of this is to be seen within the UK and the response within one of the major automotive production based areas of the country, that of the West Midlands. Volume automotive production which has been one of the mainstays of the local economy is now a thing of the past. The demise of MG Rover and its subsequent purchase by a Chinese manufacturer was probably the highest profile instance, but also simply part of a wider picture of decline. One response to this has been to try to draw together small specialist manufacturers dependent upon the automotive industry into larger groupings or guilds, cooperating not only with each other but also with the academic world in order to foster and support continuing investment in design and innovation. The Warwick

Manufacturing Group based at Warwick University is the prime example of this response. The practical question remains that of how to fund creativity and innovation when one's competitive base has been substantially weakened by overseas competition and the scale of operation that remains cannot fund high levels of risk and lengthy development times. This takes us back to the efforts of the Dyson Foundation to at least support design functions within the UK even though the actual manufacturing has moved elsewhere. What is driving this whole process is basically a very particular form of global economic policy pursued by both the developed and the fast-developing nations as they make up ground. Is there no alternative? Was it always going to move in this direction with the consequent challenges for those in the engineering profession? Is the answer simply to accept that large-scale production will shift inexorably to the Far East leaving some innovative work in the EU and the USA linked probably to companies that operate globally, plus some smaller and more networked businesses that do not have substantial overheads? If one agrees that this is the case then it would appear to be colluding with some form of economic determinism. But is it economics that has created this global environment or rather the result of particular political and strategic business decisions that favour certain vested interests? It is to this question that we must now turn.

1.6 THE RISE OF THE NEW ECONOMY

I will now examine the argument that the current context is the result of a series of decisions and policies that some within the field of politics and sociology term the rise of the New Economy. In other words, there are clear and identifiable changes within the global economic structure that derive directly from deliberately pursued political decisions and that reflect the interests of a certain set of key players. One of the major contributors to this theory is the sociologist Castells who has written one of the seminal works in recent years on the subject of what he calls the Network Society (Castells, 2000).

This New Economy has emerged during the last quarter of the twentieth century and is characterized by three distinctive features. First, it is informational in that the productivity and competitiveness of all agents in the economy (firms, regions or nations) depend upon their capacity to generate, process and apply knowledge-based information. Secondly, it is global as the core activities of production and consumption as well as their components (capital, labour, technology, markets etc.) are organized on a global scale. Finally, it is networked because it is through networks that competition is played out on the global stage. The information technology revolution has been instrumental in creating this new set of conditions and both the constraints and possibilities that flow from them (Castells, 2000, 77). What are the implications of this for the ways in which work is now organized?

Castells suggests that companies operating in this new environment have a number of strategies that they can pursue towards both skilled and unskilled labour (Castells, 2000, 254).

These are as follows:

- Downsize the firm, keeping the indispensable highly skilled labour force in the North while importing inputs from low-cost areas (very much the Dyson approach to matters).
- Subcontract part of the work to their transnational establishments and to auxiliary networks whose production can still be internalized through the network enterprise system.
- Use temporary labour, part-time workers, or informal firms as suppliers in the home country.
- Automate or relocate tasks for which the standard labour market prices are too high.
- Obtain from the labour force agreement to more stringent conditions of work and pay as a condition for the continuation of their jobs, thus reversing social contracts established under more favourable conditions for labour.

It is possible that any combination of these will be encountered in specific situations depending upon local conditions and decisions. The effect is to draw all countries into this system though and to create a convergence of labour market conditions across the globe. Furthermore, as Castells says:

> The pressure towards greater flexibility of the labour market and toward the reversal of the Welfare State in Western Europe come less from the pressures derived from East Asia than from the comparison with the United States. (Castells, 2000, 254)

Any company wishing to compete on anything like equal terms with a U.S. based business will have little choice but to follow the same route of creating greater labour flexibility. Hence "lean production, downsizing, restructuring, consolidation, and flexible management practices are induced and made possible by the intertwined impact of economic globalization and diffusion of information technologies" (Castells, 2000, 255).

So although there is not a unified global labour market, similar patterns of labour organization emerge across national boundaries. Was this inevitable though? Castells suggests not:

> This model is not the inevitable consequence of the informational paradigm but the result of an economic and political choice made by governments and companies selecting the 'low road' in the process of the transition to the new, informational economy, mainly using productivity increases for short-term profitability. These policies contrast sharply, in fact, with the possibilities of work enhancement and sustained, high productivity opened up by the transformation of the work process under the informational paradigm. (Castells, 2000, 255)

What begins to emerge from this, assuming Castells is correct, is that the way that engineering is now being shaped as a result of the forces of globalization, is only one possibility, albeit the one that leads to short-term gain for some but at the expense of others. Is this good for engineering or good for engineers? If one contrasts what is occurring with other possibilities, then one might argue that it is not. What we do have at the moment is a loss of jobs in the North combined with more stringent working conditions and the general demise of earlier victories gained by the labour movements. Increasing instability and job insecurity are combined with the downgrading of newly incorporated urban labour in industrializing countries. This is not the result of the structural logic of the informational economy which could just as easily have led to higher levels of secure employment and greater opportunities for investment and innovation. Instead, the use of networking and the political decision to create a more mobile and volatile labour market have undermined such possibilities. Now that these processes are 'locked in' it is going to be extremely difficult to reverse them.

Agencies such as the International Monetary Fund and national government organizations have suggested that the problems of rising unemployment, income inequality and social polarization are the result of a skills mismatch, exacerbated by a lack of flexibility in the labour market—we have encountered such interpretations already in the area of SET. So there is a shortage of requisite skills to enable people to take advantage of the New Economy and this is to be tackled through the educational system. However, some now argue that the evidence for this is actually extremely thin and that simply increasing the numbers of people in training will not itself create jobs that have now gone overseas. Castells documents the evidence for what has been happening in the USA in terms of unemployment and growing inequality (Castells, 2000, 298), but also points out that similar trends are visible elsewhere in the developed world. The new vulnerability of labour is not confined to low-skilled jobs but is spreading up the labour market hierarchy and into the ranks of professionals. Political parties of all persuasions on both sides of the Atlantic are pursuing the same policies with very much the same results.

> Membership of corporations, or even countries, ceased to have its privileges because stepped-up global competition kept redesigning the variable geometry of work and markets. Never was labour more central to the process of value making. But never were the workers (regardless of their skills) more vulnerable to the organization, since they had become lean individuals, farmed out in a flexible network whose whereabouts were unknown to the network itself. (Castells, 2000, 302)

So this is the wider context in which engineering is now operating, one in which instability and uncertainty have been deliberately built into the system in ways which discourage individuals from challenging the decisions and policies of their employers and which may also act as a disincentive for businesses themselves to invest in innovative or high-risk projects. Unless a project can show a swift profit turnaround for its investors—i.e. within 12 months is the figure I have heard in some companies where the real lead time for research and development would expect to be considerably longer—there is no chance of funding. The more 'competitive' the global economy becomes, the less likely it is that smaller companies will survive or that general deteriorating working conditions will be reversed.

1.7 WORK IN THE NEW ECONOMY

The picture that is emerging so far is increasingly pessimistic and points to the possibility that engineering finds itself engulfed by economic forces that threaten to stifle creativity and innovation within the profession. Stringent working conditions and lack of long-term investment by companies not prepared to buck the trend of simply acquiescing to the demands of shareholders suggest that it is going to become more difficult to recruit people who are interested in anything else than earning a living. Ideas of vision, of serving the wider public, or that engineering as a profession carries a sense of vocation seem to be losing ground. Enclosures rather than thresholds are the order of the day within this New Economy.

Before we move on to examine the detailed arguments about globalization and then to ask where and how alternatives to the current system begin to appear, we need to press more deeply into this hostile territory in order to complete the negative side of the picture. This takes us into ideas about the nature of work itself and the writings of an American sociologist, Richard Sennett. In his most recent book, *The Culture of the New Capitalism* (Sennett, 2006), he offers an interpretation of working life as it is developing within the largest U.S. and global companies. By this he means businesses with at least 9000 employees. He is clear that what he is saying about these is not yet true of many smaller companies, but that these are the path breakers and trend-setters of the new capitalism whose means of operating will influence others or prove to be the pattern of the future. They may be the tip of the iceberg at the moment, but more companies will soon be falling into line, very much as Castells also believes.

The days when companies were organized along military lines where everybody knew where they fitted into the structure, but where there was also a degree of security, stability and loyalty on both sides, are receding fast. What lies behind this? The shift from managerial to shareholder power is central to this. Sennett identifies the breakdown of the Bretton Woods agreements during the 1970s as the point when this shift became possible (Sennett, 2006, 38). This enabled wealth to move more freely around the globe and for rich nations and significant investors such as giant Pension Funds to search out potentially profitable businesses as a source of swift profit. Once this began to happen a new driving force came into play, one which had no real interest in or concern for the actual culture of an individual company, or even for the nature of the business. Only share value motivated the investment, what some have called 'impatient capital'.

The time scales involved in the investment market now began to shorten dramatically. Whereas in 1965 American Pension Funds held stock on average for 46 months, by 2000 this figure had fallen to 3.8 months. Companies came under pressure to present themselves in their best possible light to potential shareholders in this 'beauty parade' to attract investment. This often meant appearing to be dynamic and flexible rather than stable and secure. Higher staff turnover and the willingness to rationalize by cutting overhead costs by shedding staff became the signals that a business had moved into this category, whether or not such actions were really in the best longer term interests of the company.

Alongside this has run a propensity to centralize management structures and focus control at higher levels in the company. This has been aided by the implementation of more sophisticated information technology and the possibility of replacing lower level staff with automation. Why employ people when machines can do the job faster and more cheaply? As Sennett reminds us, only the larger businesses with the resources to invest are initially able to go down this route, but they are the ones which are held up as the gateway to a more profitable future.

What are the consequences of this for the nature of work itself? We have seen some of these in Castells' comments about the increase of part-time work and shorter working contracts. High mobility and flexibility become paramount. Temporary work accounts for 8% of the workforce in the USA currently (Sennett, 2006, 49). But this is only one facet of the restructuring that now characterizes such companies. Greater degrees of uncertainty, inbuilt instability and shorter term projects and agreements become the norm for staff in this environment. What are the true costs of this approach? Sennett suggests that although the financial rewards appear to be significant, what now happens is that there is a decrease of loyalty to the business, a loss of informal trust (social capital) between the staff and indeed a weakening of the institutional knowledge that often characterizes more stable businesses. People's willingness to 'go the extra mile' on behalf of the business in return for feeling that one's work is valued and appropriately rewarded is eroded. One becomes once again perhaps as in the earlier days of capitalism a 'cog in the machine', of purely instrumental value, serving the end of increasing shareholder value rather than the internal nature of the work itself.

This may have a direct impact upon the well-being of the employees.

> Pressure becomes a self-contained, deadening experience in firms with low social capital, and employees who experience pressure on these terms are far more likely to become alcoholic, to divorce, or to exhibit poor health than people working more than ten hours daily in high-loyalty firms. (Sennett, 2006, 66)

This in turn begins to impact upon the nature of the work itself, especially when this has traditionally depended upon longer time scales for its research and development functions. Placing strict limits upon the time scale within which projects have to yield a return on the investment will inevitably begin to restrict the type of project that wins funding. Either that or people will learn to 'play the game' and to present projects in ways that appear to minimize the time commitment and then hope that having once won the funding one will be able to stretch the time out if not the actual funding. Integrity rapidly becomes another victim of this type of pressure. Delayed gratification as a motivating factor is abandoned both for the investors and the work force and this impacts upon levels of commitment to the overall business. Unless rapid or immediate results can be shown nobody is allowed to gain from the work itself. This also means that the wider social networks and relationships which are often important within a research-based business are put under threat simply because groups are forced into competing between themselves as a means of obtaining the cheapest product and forcing down labour costs.

Sennett also suggests that this heralds the spectre of uselessness for many employees. How long will my role in the company last and what happens to me when I am no longer profitable or useful? Older employees more accustomed to having a secure position within the structures and therefore also more confident at challenging and questioning become a liability and tend to be replaced by younger people who now accept that 'nothing lasts for ever' and that there are no 'jobs for life' any longer, so are prepared to keep their heads down and get on with the task as given for as long as it lasts and then prepare to move on. If there is such a thing as an established corporate wisdom, then it is sacrificed on the altar of short-term shareholder profit.

What is also in danger of being lost is what Sennett calls craftsmanship. By this he means the understanding that one can derive satisfaction from being able to commit oneself thoroughly to a particular task or function and devote time and energy to it over the longer term. Pride in one's work and pleasure from having 'done a good job' no longer count for much in this new mobile and flexible environment. The skills and experience that have often been built up over a long period of time become a liability rather than an asset as the prime directive is the willingness to abandon what one has been doing and move seamlessly into another sphere of work whether or not one has the aptitude or commitment to do it justice. Skimming across the surface and giving the impression of being in command of the latest task become more important than a willingness to 'dwell' and give real consideration to a project. Everything is 'for the time being only' as Bauman puts it (Bauman, 2006, 116–118). This must have its human and business costs for those who require to give real time and attention to their work, and I assume that this applies to the more innovative and creative aspects of engineering. All in all this presents a pessimistic scenario for the future of the profession, particularly as larger companies come to dominate the field in the face of global competition.

CHAPTER 2

Who Runs Globalization?

2.1 THE ROLE OF ENGINEERING IN THE NEW ECONOMY

Having established that global forces are impinging upon the engineering profession I will now move on to consider who is actually in charge of these processes, who benefits and therefore who is likely to lose out. It is already beginning to emerge that there are difficulties in terms of working conditions and the prospects for both investment and innovation within the field. So globalization—and I will examine the term itself in more detail in the next chapter—might not be good for engineers. But there is, I would suggest, a wider issue of whether or not globalization is good for other people. If it were to become clear that it is not, should this be of concern to engineers? Does it matter if there are people or groups losing out as a result of what is happening and is it the role of engineers to express a view on this or to be uneasy about it?

I would suggest that it does matter to engineers and that it is not good enough to shrug one's shoulders and argue that this is life and that one simply has to get on with the work regardless of the consequences as long as the profits are still rolling in. Part of what it means to be a 'professional' is surely that it is not just a matter of possessing a particular expertise but that it also involves standards of behaviour and codes of practice. Other texts in the series will elaborate upon the exact content of these and there may well be legitimate differences of opinion over the detail, but the principle stands that there is more to being an engineer than simply exchanging one's professional skills for an appropriate remuneration. Questions of values, of motivation, of due concern for one's customers let alone the wider community and even the planet itself need to be on the agenda if professional standards are to be protected. Hence the further question of who benefits from this particular brand of economically driven globalization cannot be ignored.

If it is the case that certain groups are shaping and then exploiting current global development for their own ends, and, again, the detailed arguments will be considered in the following chapters, then others who are complicit in this need to be challenged. It could be that engineers are allowing themselves to be appropriated by other agendas derived, for instance, from the drive for short-term shareholder profit and are thereby colluding with a neo-liberal project that can only succeed at the expense of poorer groups around the globe. If this is not the case, then it is important to know that and to be able to articulate alternative approaches. If there is a real danger that this is happening by default, as engineers (and others of course) fail to speak out against or even to become aware of the consequences of economic and political decisions that are being made, then engineers as responsible moral agents might need help in understanding and then questioning the powerful forces that are shaping globalization. The identity and integrity of the engineering profession require a critical public stance over issues of practical and moral concern. Is globalization putting the integrity of the engineering profession at risk and will the identity of its tradition of working for the well-being of others be undermined by purely financial considerations? If such threats are real, then this is about so much more than simply the working conditions of engineers themselves, although glimpsing how those are being threatened is a clear way into the wider debate. Scientific objectivity must not be used as a shield for questionable political practice. Research, creativity and innovation should be thresholds into new and beneficial possibilities. If they are more likely to form the substance of new enclosures instead, then close scrutiny needs to be exercised.

2.2 CHINA OR THE USA?

So who is controlling the processes of economic globalization? It is important that we get a grasp of the current situation and the possible future directions for development. Turning first of all to the engineering profession itself, one can rapidly see that China is starting to challenge the former dominance of the USA. One major U.S. company, GE plastics, is now well on the way to selling $1 billion's worth of advanced materials in China over the next two years (Fishman, 2005, 216). Manufacturing is booming and there is a clamour for places in higher education by budding engineers and scientists. GE has already opened a giant industrial research centre in Shanghai and this year expects to employ 1200 people in its Chinese laboratories. It has also set up scholarships at leading Chinese technical universities.

This type of expansionary programme is getting internal support from the Chinese government. The country already has 17 million university and advanced vocational students (up more than threefold within 5 years), the majority of whom are in science and engineering. They quote a figure of 325,000 engineers being produced in 2005, which is five times the number in the USA where the number of engineering graduates has been declining since the early 1980s. Even more worrying, 40% of students who enter the engineering track at U.S. universities change their mind. Even so, the current gap between the USA and China in terms of research resources remains substantial. So, for instance, the U.S. government authorized $3.7 billion to finance research into nanotechnology in 2004 and China simply does not have the funding nor the infrastructure to support programmes on this scale.

However, the gap is beginning to narrow. In more mainstream applied technology and innovation, China spent $60 billion on research in 2004. The only countries spending more were the USA and Japan, with a figure of $282 billion and $104 billion respectively. The USA still has twice as many researchers as China (1.3 million compared to 743,000), but this figure is clearly set to move in favour of China over the coming years. As we have seen from earlier examples, the pattern for now might well be that production is shifted east for obvious reasons, but that more sophisticated research facilities remain in the USA and Europe, but how long will this last as China speeds further down the road of industrial development?

Another Chinese example comes from the city of Chengdu, capital of Sichuan province in South West China. This covers an area slightly larger than California but three times as populous (Fishman, 2005, 218). So there are about 107 million people in the province, 43 universities and 1.2 million scientists and engineers. At the moment its fragmented transportation system prevents it rivalling cities further east as a manufacturing centre, but it is promoting its plentiful, relatively low-cost intellectual expertise through its new research corridor, the West High-Tech Zone. Motorola has built its newest research centre in the province and regards this as a world centre for software engineering. Other major enterprises already have bases within the province, such as Intel, Ericsson, Siemens, Alcatel and Fuji Heavy Industries of Japan. This is part of a wider picture in which between two and four hundred foreign companies have established research centres in China since 1990. This is not just about tax incentives and lower labour costs, but also related to the growing consumer base in the country. As Fishman says though, the probable outcome of this continued growth (China's aim is a year on year growth of 8% in its GDP and this is being achieved) is a saturation of the markets and concerning levels of overcapacity. While China catches up and other developed nations stand still or decline, questions of economic and political conflict will come to the fore. Is this level of growth sustainable and what are the costs both for China itself let alone the rest of the world? How far will the USA allow this process to progress before it flexes its muscles in order to protect its own market share and economic dominance?

The actual figures of the loss of jobs from the USA to China are difficult to assess as the U.S. government does not collate them, but one study in 2004 showed that during the course of 3 months, 58 U.S. companies, 55 European companies and 33 from other Asian countries, all announced plans to move jobs to China (Fishman, 2005, 273). This was a marked increase for the USA where in 2001 only 25 companies announced shifts to China. The study concludes that U.S. work sites moved 400,000 jobs to other countries over the course of 2004, twice the number that had gone 3 years before. Another significant change is the actual nature of the posts that have gone to China. Whereas they used to be in the fields of electronics and toys, now they are across a broader section of the employment pool. This is all part of the growing drive for labour flexibility that is central to the New Economy. Local labour forces will have less bargaining power as the threat is always present that jobs will simply be moved to other (cheaper) countries if terms favourable to management are not accepted.

Another implication of these changes is the possibility that companies who retain research facilities in the USA and Europe will find that they reach the point where they no longer have the critical mass of people to remain innovative and competitive. Where will the global entrepreneurs and innovators be located in years to come? As Fishman says:

> One of China's most potent economic weapons is its ability to attract entire industry clusters, acquiring the critical masses of companies that catalyze the creative ferment that leads to rapid innovation. Global telecommunications and regular air links may go a long way to closing the distances for Cox's American army of global entrepreneurs, but Americans stringing together opportunities in distant lands will have to spend a lot of time re-creating the network of relationships that has been lost as America's industrial clusters depopulate, devolve, or both. (Fishman, 2005, 276)

So who exactly will be in control of the process of economic globalization in a few years time and what might be the implications for engineers and indeed scientific research if the balance of power shifts decisively to the east as it may well do? It is possible of course that these arguments are being exaggerated. It may be that China (and India) are experiencing their own industrial revolutions and simply playing catch up with the West and that matters will even themselves out over time. It may also be the case that such rapid rates of growth are not sustainable and that internal political factors will start to drag China back as instability and uncertainty sets in as it already has elsewhere. However, when one considers that China and India between them possess a third of the world's population, it is clear that their economic expansion still has a long way to go and must have a dramatic impact upon the rest of the world if it continues along the current path. Although the USA still dominates the global economy, this picture is changing as we read.

2.3 LARGE OR SMALL: SHORT-TERM OR LONG-TERM?

We have already touched upon the question of the scale of enterprise that is most likely to survive and thrive within the New Economy, but it does need to be emphasized that there are significant questions here about who is going to benefit from economic globalization in its current form and that engineers need to be alert to these issues. So here is an example from Canada that sends out warning signals (Bauman, 2006, 71).

This story was reported in a regular column called 'Countryside Commentary' in the newspaper *Corner Post* on 24 May 2002 in an article written by the strategic policy advisor to the Christian Farmers Federation of Ontario. The article was entitled 'The collateral damage from Globalization'. The author reported that each year more food is produced employing fewer people and with a more prudent use of resources. So in the 4 years up to February 2002,

35,000 workers disappeared from Ontario's farming statistics, made redundant by technological progress and replaced by new and labour saving technology. However, whilst one would have anticipated that the subsequent increases in productivity should have made rural Ontario richer and farmers' profits would have soared, there has been no sign of this. The conclusion can only be that the profits have accrued elsewhere in the system.

The apparent explanation for this is globalization to the extent that it has spawned a series of mergers and buyouts by the firms that supply farm inputs. This is justified by the argument about remaining globally competitive through creating larger and possibly monopolistic groupings of companies. It is these businesses though that then reap the benefit of higher productivity. The article concludes:

> Large corporations become predatory giants and then capture markets. They can—and do—use economic power to get what they want from the countryside. Voluntary exchanges, trading goods between equals, are giving way to a command-and-control countryside economy.

Again, one needs to be a little wary of such comments as they are clearly written from the perspective of the local farming community and represent a particular vested interest. Yet they are consistent with what is evident within other sectors of U.S. and European economies. What is happening is that globalization is being used as a reason for concentrating power within fewer businesses on the grounds that this is the only way to remain competitive given the growth of China, etc. Only the larger companies who have economies of scale will be able to survive and compete. The alternative to this is a growth in smaller and more niche marketed or specialized companies who rely more on networks and IT than on substantial plants or work forces. Volume production in most industries has moved east. One needs to question whether this movement is inevitable or whether the argument is sometimes presented as a justification for decisions that are really being made for other reasons. Who gains from these decisions and who loses out? Is the name of the game that of chasing short-term profit whatever the longer-term local costs? It is easier for companies to transfer production to the other side of the world—or it might be—than for the work force with its local commitments to even consider such a move. But the whole argument rests on the assumption that 'in the highly competitive global economy big is beautiful once again'. One wonders on what grounds such a view can be supported and justified.

Yet if 'large versus small' is a dimension of the current debate about globalization, then there is clearly also a direct link to questions of time scale and subsequent concerns about stability and security for projects and employees. The New Economy demands flexibility and the capacity to shift from one set of skills to another at almost a moment's notice, whatever the personal costs. Bauman presents a powerful analogy that is worth noting (Bauman, 2006, 117–118). He likens this process to the contrast between bullets fired from a ballistic weapon and the new smart missiles. The trajectory and direction of the bullet are predetermined making them ideal weapons in positional warfare where targets remain static and predictable. However, once targets become more volatile and unpredictable in their movements, bullets become ineffective and smart missiles that can themselves change course and track the target are what is required.

> Such smart missiles cannot suspend the gathering and processing of information as it travels, let alone finish them—its target never stops moving and changing direction and speed, so that plotting the place of encounter needs to be constantly updated and corrected. (Bauman, 2006, 117)

This is a way of describing the 'instrumental rationality' that now appears to dominate current business culture. In other words, it is no longer simple to calculate the ends for which a particular weapon is the means of destruction. The target itself may change en route and the missiles will be programmed to choose the target once they are in flight. So some sort of general capacity to adapt and then pursue a goal that might change as one moves is of more value than the old ability to pursue a pre-identified and predetermined goal. The missile must learn as it goes along and what matters is the ability to learn fast, forgetting what must now be abandoned and ignoring what might have claimed to be important information when the journey began. So knowledge itself becomes disposable: useful 'for the time being only' and to be dumped as soon as the target changes. Everything is good only 'until further notice'.

This has considerable implications for teaching and training which used to be based on the assumption that imparting certain skills and knowledge was required in order to equip people for predictable and identifiable tasks. If this has now changed—and one can recognize enough of Bauman's descriptions of 'liquid life' to see how this might be the case—then it will have an effect upon how people have to be prepared for their working lives even within the fields of engineering and scientific research. It will not do any longer to train people to follow a specific career path or a subdiscipline within a profession because that will leave them 'high and dry' when the task changes and the target has shifted. It is the more general capacity to let go of what appeared to be important in order to adopt what is now important that is required. The question though is that of the depth of knowledge and understanding that will be acquired once this more 'superficial' approach begins to take over. Is it a matter of knowing a little about everything or just enough to bluff one's way through the labyrinth rather than of accumulating the knowledge which comes from time-bound experience within a particular field? If teaching and training themselves become so geared to these requirements of the New Economy what standards will remain? Is such short-termism really of benefit to skills acquisition and building up a high-quality work force with an appropriate depth of knowledge?

Perhaps Bauman is overstating the case and such worries are exaggerated. But perhaps he has a point, in which case the arguments about globalization used as a justification for both large-scale production and research facilities dominated by global companies and then smaller and smaller time-scale approaches to skills and working practices need to be questioned. Engineers drawn into this process require perhaps reflexivity rather than flexibility!

2.4 NATION STATES OR GLOBAL BUSINESS?

So who is 'running the show' and who benefits from globalization? The exact balance between the power of nation states and that of companies will form a focus for the next chapter, but it is worth at this stage looking at how companies now operate in the New Economy and who does 'call the shots'. Is it owners, senior managers (CEOs) or shareholders who now determine what is to happen? According the French sociologist Bourdieu it is none of these (Bourdieu, 2003, 28).

> It is in fact, the managers of the big institutions, the pension funds, the big insurance companies, and, particularly in the United States, the money market funds or mutual funds who today dominate the field of financial capital, within which financial capital is both stake and weapon. These managers possess a formidable capacity to pressure both firms and states.

Their bargaining strength means that they can effectively impose upon others what is termed a 'minimum guaranteed shareholder income'. Thus ever higher profits become the immediate and short-term goal of any business and are often achieved through downsizing and labour reductions. Bourdieu is close to suggesting that this is some sort of infernal machine that operates according to its own internal logic almost behind the backs of the people apparently running the system. Stock market value becomes the only measure of success and if values fall then the system demands it own 'victims' as some sort of sacrificial lambs.

> Thus has come into being an economic regime that is inseparable from a political regime, a mode of production that entails a mode of domination based on the institution of insecurity, domination through precariousness: a deregulated financial market fosters a deregulated labour market and thereby the casualization of labour that cows workers into submission. (Bourdieu, 2003, 29)

The New Economy is dominated by individuals who are international rather than belonging to any particular state, polyglot and polycultural. It is immaterial and 'weightless' in that it circulates information and cultural products rather than consumer durables. In other words, it belongs to a cultural elite who appear to form an 'economy of intelligence'. By contrast those who fail to benefit from the advances that are being made are somehow lacking in the required skills and intelligence to work out how to play this new system. This is a new form of the survival of the fittest. So it not straightforwardly the case that companies themselves are the immediate and obvious beneficiaries of globalization—indeed businesses come and go, are reconfigured and swallowed up in the process of mergers and acquisitions—but those who learn to 'travel light' and shift their investments rapidly and cleverly are the gainers in this system.

Bourdieu responds to this by suggesting that new forms of struggle must be developed. Given that 'ideas' and 'knowledge' are central to the New Economy, albeit in constantly shifting forms, researchers themselves have a critical role to play. It is up to them to provide alternatives to the current system in ways that others can appropriate and access. This must transcend national boundaries though if it is to have any impact, drawing together diverse strands of protest and opposition and forming new collectivities and alliances. Suggestions for alternatives to the system will emerge in the final chapter, but it is important to note the various arguments about who gains and who loses in the game of globalization and what the effects of this are likely to be on the engineering profession.

We return then to an earlier question about the role of engineering within this New Economy. Are engineers and scientific researchers merely 'assets' to be employed, made redundant, redeployed if they are lucky or rather left on the scrap heap as financial capital decides to move on ever eastwards and upwards? The sorts of arguments presented so far suggest that this is likely to be the case. Even debates about the value of 'human capital' and the skills and knowledge possessed by such groups appear to carry little weight within this sort of economic configuration. In fact, such 'solid' and 'reliable' characteristics could even be seen as a liability rather than an asset if Bauman and Bourdieu are correct. But then one must question whether it is possible for true innovation and creativity to flourish in this sort of a working environment where the institutionalization of precariousness and inbuilt instability are determining factors. Can engineers give of their best when so much is uncertain and constantly shifting? Will others then be enabled to benefit from their skills and expertise? If globalization means exclusively this particular form of economic culture, then it appears to be an enclosure of a really damaging nature to the profession. But is this the only way of interpreting globalization and are there truly no alternatives? It is to these questions that we turn in the final two chapters.

CHAPTER 3

How to Interpret Globalization

3.1 INTRODUCTION

The point has come to expand the discussion beyond the immediate impacts of globalization upon the engineering profession and to examine how the term itself is used and what might be learnt from that. So far what we have encountered are a series of enclosures that are already affecting engineers derived from changes in working practices and increased competition from China and other rapidly developing nations. Yet I have tried to emphasize that the New Economy is only one way in which globalization might be shaped, that things do not have to be this way and the fact that they are is the direct result of political and economic decisions made by particular parties who believe they will make significant financial gains from this configuration.

For brief confirmation of this I turn to Joseph Stiglitz, Chief Economist at the World Bank until 2000 and before that Chairman of President Clinton's Council of Economic Advisors. His recent publications on the subject of globalization reveal the different interpretations of what has been happening in terms of global economic development (Stiglitz, 2002, 2003). His major concern is that economic and financial factors have been at the heart of these developments but at the expense of wider social and political issues. As a result of this there are growing inequalities both within societies and between nations, unemployment is starting to impinge on Western nations once again and the possible advantages of a global economy are being squandered because of a concentration on short-term profit. He says:

> Globalization affects the kinds of societies that are being created throughout the world. And it is precisely because those in the rest of the world are aware of this that emotions about globalization run so high: we have been pushing a set of policies that is increasing inequality abroad and, in some cases, undermining traditional institutions. There is an alternative vision, one based on global social justice and a balanced role for the government and the market. It is for that vision that we should be striving. (Stiglitz, 2003, 319)

Whilst I generally agree with that I also want to suggest that what is required is a broader understanding of what is meant by globalization. We need to move beyond the narrowly economic to get a grasp of the other dimensions of what is now happening and that fall under this heading. Therefore in this chapter I will examine a variety of interpretations which will offer a glimpse into these and also present ways in which thresholds into new possibilities exist beneath the surface of the current concentration upon the New Economy and its problems. Is what is being called globalization better referred to as globalism for instance? So this refers solely to the advance of neo-liberal deregulatory economic policies being driven by the 'Washington consensus'. Then globalization would refer to a wider set of social processes touching on issues such as culture and identity and ways in which groups and individuals respond to the blurring of boundaries that now occurs at an increased frequency. Where are the sources of potential opposition to the current regimes and how effective are they likely to be? How are people of moral conviction and integrity who have that concern for social justice to begin to exercise that? To help identify potential thresholds as they might emerge for the engineering profession we turn first to definitions of globalization and then examine other issues that need to be highlighted.

3.2 HOW TO DEFINE GLOBALIZATION

Globalization is "the expanding scale, growing magnitude, speeding up and deepening impact of interregional flows and patterns of social interaction. It refers to a shift or transformation in the scale of human organization that links distant communities and expands the reach of power relations across the world's major regions and continents"—but it is not a harmonious process as it leads to conflicts and divisions as not everybody is part of it (Held and McGrew, 2003, 4).

An important debate is whether or not this is actually anything new, but rather a continuation of a process that has been occurring over a long period of time. There are those (e.g. Beck) who point out that there have long been degrees of what is now described as globalization, but that what we see now is an intensification of this process combined with a growing self-awareness of what is happening and its impact upon our lives (reflexivity). Various versions of the term cosmopolitanization are preferred for this while the term globalization refers more strictly to the political process of economic liberalization associated with Western government policy and such institutions as the World Trade Organization (WTO), World Bank and the International Monetary Fund (IMF).

3.3 MAJOR ISSUES

Much controversy also surrounds the role of the nation state in this process. How much autonomy does a typical nation state now possess in the face of the supposed global forces that determine our lives? What about the power of transnational flows of capital, transnational corporations whose turnover might well exceed the gross national product (GNP) of many smaller countries, let alone international non-governmental organizations (INGOs) which are often the main sources of protest? Then there are the risks that are now part of globalization and which are no respecters of national boundaries, e.g. environmental, nuclear accidents, international terrorism, etc. One might reasonably expect that engineers have a significant contribution to make to the potential solutions to such problems.

Some parts of the world are also engaged in greater regionalization and this too might undermine the power and sovereignty of the nation state. Global culture and the influence of the global media are a further factor and might be seen to lead to a growing standardization across national and cultural boundaries, although they can also have the impact of heightening differences as people become more aware of the need to clearly identify what is specific to them. 'Traditions' are revived in a self-conscious and 'non-traditional' manner, either to support a sense of national identity in the face of such standardization or as a marketing ploy in the competition to attract international tourists. Even companies deliberately 'adopt' supposedly local practices and customs in order to contextualize their products and processes.

Is there a growing global civil society? Once again this is matter of controversy. Some see the anti-globalization movements, the influence of INGOs and the ease of access to the internet and global travel and communications generally as means by which a global civil society is coming into being. Others are more sceptical and argue against any coherent or structured movements that could in any way be representative, particularly when huge swathes of less developed countries are still excluded from the economic aspects of globalization. Is this a realistic or even desirable aspiration? Is globalization really restricted to an elite, either economic, political or academic, who take for granted the international contacts that now make up their working lives but are an exception rather than a rule for many others?

Finally, is globalization simply a thin disguise for the political and military dominance of the USA particularly since the end of the Cold War and the apparent demise of communism as the only extant alternative to Western (global) capitalism? Whose interests are served by these processes and whose policies really determine what is happening in the global economy? Where does the European Union stand in this and what account does this take of the growing power of China, India and other nations in the Far East who might soon be in a position to challenge the hegemony of the USA, if they are not already doing so? What is the role of the United Nations, the Security Council and other transnational governmental or governance-based bodies? Are they simply 'pawns in the American game' or do they have a degree of autonomy and authority?

This suggests that 'globalization' is itself a highly contested concept that contains a number of quite different strands and arguments and that must therefore be treated with a degree of academic rigour and indeed caution.

3.4 THE MYTH OF GLOBALIZATION

Here is an example of an argument against the concept of globalization mounted by Hirst and Thompson.

Globalization (in its radical form) is a myth for the following reasons:

- The present highly internationalized economy is not unprecedented and in some ways it is less open and integrated than the world between 1870 and 1914.
- Genuine transnational companies are relatively rare as most are still based on a major national location.
- Capital mobility is not producing a massive shift of investment and employment from the advanced to the developing countries.
- The world economy is far from being genuinely global as it is dominated by the triad of Europe, Japan and North America.
- These three areas exert considerable influence and governmental pressures over the rest of the world.

Much of this might now be contested but it does illustrate that the theories of globalization are not universally accepted or interpreted within the fields of economics, sociology or politics.

There are further levels of this debate which take us beyond those disciplines however and into the territory of intellectual history. Thus globalization is seen by some as a further stage in human development that can be traced back to the Enlightenment. Within the latter there were three big ideas: there is a third world; it is possible to imagine and work towards a better world—i.e. the idea of human progress; this progress will be achieved by and through the nation state. Globalization has brought each of these assumptions into question.

Thus the 'third world' is now to be found in every country not simply in certain areas of the world. Each nation now has its excluded sections, those who do not share in the benefits of economic globalization. So there are divisions within countries and between different regions within countries. Whilst the politics of inequality or rather the struggles against it, used to dominate political thinking on the left, this has now been undermined by the politics of equality and diversity, where 'difference' is seen as a positive and desirable good. This is a highly significant and controversial debate to which faith groups contribute in various ways.

3.5 THE IDEA OF A RISK SOCIETY

The idea of progress has now been replaced by that of risk and the tasks of the management thereof. So we refer once again to the notion that environmental let alone economic and political dangers are no respecters of national boundaries, but now also intensified by the awareness that our investment in 'progress' brings unintended and unforeseen consequences that potentially undermine the positive aspects of change. Ulrich Beck is probably the major thinker who has brought this to the surface (Beck, 1992). An example would be the greater risk of the rapid spread of diseases because of greater international travel. There are interesting debates here about the extent to which national governments explicitly use fear of such 'global threats' as a means of control of their populations, particularly since the end of the Cold War and the demise of a military 'external other'. However, there often appears to be a prevailing air of pessimism about the future which challenges the apparently deeply embedded idea of progress and unimpeded development and growth.

The role of the nation state in all of this is clearly central, although its powers to respond to these new challenges are now in question. The problem is that of whether if it is true that its powers have diminished, enough of a global or international infrastructure has yet developed which can safeguard both national and international security. Many would argue not and continue to present the view that the interests of the major economic powers are pursued at the expense of the lesser countries. In which case globalization will simply exacerbate current inequalities and lead to further potential conflict. So globalization is seen as basically an unjust system. Is there any source of hope for a reversal of this? One major candidate for this role is the counter-arguments of environmental groups and ideas! Some believe it is already too late for this to have any effect, others are more optimistic.

3.6 CRITICAL QUESTIONS FOR ENGINEERS

What are the questions that we might usefully ask here? First it would be good to get a sense of the strength and weakness of the economic arguments about globalization. However, a lot will depend on which economists one feels able to trust and it is difficult for non-experts to know how to assess the various arguments. In general terms, it does appear pretty clear that significant changes are taking place on a more than national level and these impact upon our working lives—so global competition for markets impacts national unemployment, the vibrancy or otherwise of local economies let alone terms and conditions of employment. Longer working hours and decreased power of unions are familiar aspects of the UK economy and are often justified by the argument that these are needed to remain competitive in a global economy.

Second is the more philosophical debate about the notion of progress which may or may not be attributed to the Enlightenment. Certainly arguments about post-modernity in the 1990s talked a great deal about the breakdown of the meta-narrative of the Enlightenment and some saw this as an opportunity for religious narratives, for instance, to reassert their dominance. However, whether or not one goes that far, we still have to ask whether it is true that the idea of progress has lost its influence to quite that extent. One might say that the philosophy which underlies science, engineering and technology is based upon the Enlightenment idea of progress through the exercise of human reason and autonomy, so if this is now under challenge engineers might have to re-think their sense of identity and role within a different sort of world.

How does one approach the problems of 'third world' countries or areas within countries that still lack basic amenities and that might be held back if environmental regulations are imposed by the developed world? Questions of justice are surely paramount here and cannot simply be dismissed by an abandonment of ideas of progress and development. What about our reliance upon scientific research, some of which is based upon the defence industry for sure, but some is also geared towards the improvement of human life and the results and gains of which many of us take for granted? Yes, we know there are risks involved and many of us no longer have an unquestioning confidence in 'scientific experts', but this is not the same as totally abandoning any idea of human progress!

Then there is the vexed question of how much power nation states do still possess and how protest groups relate to and also critique governments and international governance bodies. Are we now all 'inside' the empire which is Western Global Capitalism as dominated by the USA and a few others, trapped in a system for which there is no alternative? The only question then is where one is going to be located within that system, as either 'winner' or 'loser', given its fragility in the face of other threats, environmental, political and economic.

3.7 GLOBALIZATION OR COSMOPOLITANIZATION?

For a helpful summary of the type of argument encountered here we turn to the work of Ulrich Beck (Beck, 2006). He states that in public discourse the now fashionable term 'globalization' is understood primarily as economic globalization and might better be termed 'globalism'. This promotes the idea of the global market, defends the virtues of neo-liberal economic growth and the utility of allowing capital, commodities and labour to travel freely across borders. It is claimed that it is this process that has led to economic growth across the globe in the last two decades, largely as a result of the deregulation of markets since the 1980s. Even opposition to this process rests upon the same assumptions to the extent that it defends the power of the nation state against global economic forces.

Cosmopolitanization however is a multi-dimensional process which goes well beyond the purely economic and involves politics, culture, civil society and developing international bodies and structures. Yet this process has been underway for many generations, ever since boundaries have been more regularly crossed by increasing numbers of people. This blurring of boundaries can be seen in what Beck calls 'banal cosmopolitanism' and raises interesting questions of hybridity and purity for all cultures and traditions, including those of science and engineering. He offers as an example the modern odyssey of 'authentic' Indian cuisine.

> Anyone who thinks that the trademark 'Indian restaurant' implies that Indian cuisine comes from India is sorely mistaken. Indians in India have no tradition of public restaurants.... The Indian restaurant is an invention of Bengalis living in London, as are the 'exotic dishes' which are now celebrated and consumed all over the world as ambassadors of Indian traditions. In the course of its march to globalization, the Indian restaurant and its characteristic menu were also ultimately exported to India, which stimulated Indian households to cook Indian food in accordance with London inventions. Thus it came to pass that today one can eat 'Indian' food even in India, thus confirming the myth of origins. (Beck, 2006, 10)

Hence the main manifestation of this process is the blurring of boundaries, the mixing of traditions and ideas and the intermingling of elements of different cultures and lifestyles. This overlapping of places and times creates ever new 'salads' and combinations that encounter, depart, re-encounter in new forms and are constantly in a state of flux and movement. None of which prevents us from trying to 'tie things down' and keep them still in order to provide us with some element of stability and security.

Is this cause for celebration and optimism as we enter a 'brave new world' where differences and divisions based on those old traditions are consigned to the dustbins of history? Beck says not:

> It does not herald the first rays of universal brotherly love among peoples, or the dawn of the world republic, or a free-floating global outlook, or compulsory xenophilia. Nor is cosmopolitanism a kind of supplement that is supposed to replace nationalism and provincialism, for the very good reason that the ideas of human rights and democracy need a national base. Rather, the cosmopolitan outlook means that, in a world of global crises and dangers produced by civilization, the old differentiations between internal and external, national and international, us and them, lose their validity and a new cosmopolitan realism becomes essential to survival. (Beck, 2006, 14)

What we encounter then are increasing confusion and complexity, within which what we thought we knew can rapidly dissolve into new combinations and frightening possibilities. A retreat to tradition, and here we might include the self-understanding of engineering itself, can be a reaction against this and part of the search for security and stability. Are new approaches such as a growing support for environmental concerns simply part of this picture, a clinging to the life raft or the threads of what now remain from older traditions combined with a renewed sensitivity towards non-human life and worlds? Or is there a real hope for the formation of a new and as yet undeveloped human identity which might be less destructive of itself and others?

3.8 THE IDEA OF A GLOBAL COVENANT AND THE ROLE OF GLOBAL GOVERNANCE

One of the leading academics working in this field is David Held, whose work has already been drawn upon. He stands broadly within what might be termed a humanist–rationalist approach to political sociology and sets considerable store by the prospect of international structures that could effectively control and/or guard against the worst excesses of globalization. In particular, he believes that it is possible to present an alternative to what is often termed 'the Washington consensus', by which he means the processes of trade liberalization and openness to market forces that is invariably identified with an American right wing government political stance.

In one of his most recent texts (Held, 2004), Held lays out what he believes to be some of the current myths about globalization. Globalization does not equal Americanization, and is not to be identified with Western imperialism. Which is not to say that both are not important influences in the process, but it needs to be noted that U.S. companies only account for one fifth of world total imports and about one quarter of total exports. So globalization is not just an American phenomenon. Neither must it be assumed that it means a 'race to the bottom' in welfare and labour standards, and Held notes the diversity of welfare models that operate across Europe and ways in which some governments maintain strong welfare regimes. However, once again, there are counter-examples of this in developing countries that do not possess an appropriate political and social infrastructure. Then, he claims, there has not been a simple collapse of environmental standards, even though it is clear that much remains unaddressed and unresolved in this area. But he sees these problems as political and ethical rather being the direct result of globalization as such.

Globalization is not associated with the end of the nation state—we have seen earlier that this is one of the great fears emerging from the critics of globalization—but nation states remain the primary actors in world affairs according to Held. Their power has probably been reconfigured and reshaped as political complexity has increased, but is still the single most significant component in global politics. Nor is it the case that globalization has merely compounded global inequalities, and Held quotes various figures to support this argument. So although the gap between average incomes in the wealthiest and poorest countries is now further apart than ever before, the proportion of those who live in the very poorest conditions appears to have declined across the world. He acknowledges that there are variations within countries and offers China and India as examples of this.

Corporate power has not simply been reinforced by globalization; global companies may be bigger than ever before but not necessarily more powerful as they are constrained by external factors over which they have little or no control. It is not even the case that developing countries are losing out in terms of their proportion of world trade, although this rather depends on including Far Eastern nations as developing countries alongside more obvious candidates in the

Middle East and Africa. Held also maintains that such developing nations do have a say in the processes of global governance and that there are grounds for being optimistic about the possibilities of structures of global civil society being the engine for positive change (Held, 2004, 3–10).

The general drift of the argument is clear. Although there are indeed major challenges and problems associated with globalization, it is too simplistic and deterministic to state that 'things are only getting worse' for the less powerful nations and people and that they will continue to do so. The task is to have confidence in what can be achieved on a developmental model that is different from that of the Washington Consensus and to put effort and resources into making things happen rather than simply complaining about how bad things are. Much of Held's text is then devoted to how structures of global governance can be put into place or built upon in such a way that improvements might be achieved, through economics, politics and law, and, after a summary of these (Held, 2004, 164–165), he states what are his core principles or cosmopolitan values:

- equal worth and dignity
- active agency
- personal responsibility and accountability
- consent
- collective decision making about public matters through voting procedures
- inclusiveness and subsidiarity
- avoidance of serious harm
- sustainability (Held, 2004, 171).

Whilst one might not share Held's optimism about the outcomes of current trends in globalization, he does throw down the gauntlet to those 'prophets of doom' who simply present a breakdown of identity, culture and political order as the inevitable future. If the worst is to be avoided is it not up to men and women of integrity and good faith to work together in positive ways to avert these potential catastrophes and improve the human lot wherever possible? Do those who constantly warn of impending doom from climate change, energy crises and shortage of basic resources not run the risk of creating an attitude of paralyzed acceptance of the worst instead of encouraging positive action? At least Held is prepared to offer alternatives to both the doomsday scenarios and the Washington consensus.

An inescapable question for critics of globalization is that of the extent to which we can put our trust and confidence in the purely human capacity for reason and sound judgement upon which such an approach as Held's must rely. Or do we believe that it is the exercise of human autonomy that has brought us into this dangerous position in the first place? In which case, do we now abandon this route altogether as leading only to further disaster (which is what some environmentalists appear to support), or is it possible to move through and beyond autonomy to another way of being that builds upon the best of autonomy but leaves its worst excesses behind? The question of what it might mean to be a human being is surely at the very core of all the concerns emerging from globalization.

3.9 IS THERE A GLOBAL CIVIL SOCIETY?

Professionals have a natural interest in the answer to this question since they are often located in the sphere of civil society, somewhere in the realm between government and private individual and family. If it is the case that some form of global civil society is beginning to emerge, then one might expect professional groups to be part of this or, if they are not, to ask why and what might be done about it. Yet, as with each of the other areas we have looked at, the evidence is far from clear and the arguments hotly debated. I will draw here upon the work of John Keane, a long-established writer in this field who has now turned his attention to the global dimensions of the discussion (Keane, 2003).

Keane argues that there are seven converging strands that now encourage research into this area:

- the revival of the old language of civil society following the military demise of communism during the 'Prague spring';
- the increasing appreciation of the effects of global media and communications;
- a new awareness stimulated by the peace and environmental movements that we are all members of a fragile and potentially self-destructive world system;
- the possibility that the implosion of Soviet style communism might create a 'new world order';
- the worldwide growth associated with neo-liberal economic policies based on market capitalism;
- disillusion with the broken and unfulfilled promises of post-colonial states;
- dangerous vacuums created by the collapse of empires and states resulting in civil wars (Keane, 2003, 1–2).

However, none of those are conclusive proof that a global civil society is actually developing. As Keane says, the very concept can be used either to describe a state of affairs that already exists or to advocate policies that are not yet in place but might lead to this state of affairs if followed through. One of the obvious problems is that of collecting empirical data which, up until now, has been based largely on national boundaries, whereas what is required is data of such things as the flows of people across those very boundaries. The inevitable blurring that would provide evidence makes the hard facts and figures extremely elusive. Keane suggests that "global civil society is never a fixed entity but always a temporary assembly; a self-reflexive dynamism marked by innovation, conflict, compromise and consensus" (Keane, 2003, 7). He is in tune with Beck's comments about the essentially messy and difficult to define nature of what we are trying to discuss. Hence any data presented will tend to make the picture look clearer than it is and any concept of global civil society will appear abstract and distanced from the complexities on ground.

So we are dealing here with what sociologists call an 'ideal type', a dynamic system of interconnected institutions that penetrate into national territory, an unfinished and unpredictable network with hub and spoke clusters that shift backwards and forwards across national borders, consisting of groups, individuals, voluntary organizations, charities, academics, businesses, internet networks, trade unions and even religious groupings.

What is being talked about then is less some sort of coherent and identifiable 'community' but rather loose associations of people and groups which change over time, but built upon heightened self-awareness (reflexivity) which has an understanding of the hybridity and complexity of our world. Within this there are still 'no go areas' just as much as there are new and creative groupings and possibilities for shared interests and activities. There is still the danger that those involved form some sort of elite, perhaps a western European middle class, liberally minded and academically orientated or well-travelled section of the populace who are excellent at articulating their own experience but far from representative of the majority in any one country. One would need to see evidence that those involved in global activity were emerging from a wider cross-section of the population. However, Keane reminds us that much of what we talk about under this heading was around before the First World War in the guise of mapping, exploration, travel and, of course, the influence of missionary activity.

3.10 NON-GOVERNMENTAL ORGANIZATIONS AND CIVIL SOCIETY

One possible source of support for the existence of a global civil society would come from the figures relating to Non-governmental organizations and these are indeed significant. Numbers now engaged in the World Wide Fund for Nature (WWFN) and Amnesty International, for instance, exceed those for governments, the media and big business. Transnational advocacy NGOs (TANGOs) such as WWFN now have 4.7 million members in 31 countries. Friends of the Earth has 1 million members in 56 countries. Amnesty International has 1.1 million members in 150 countries (Keane, 2003, 58). Even the anti-globalization protests and recent marches against the Iraq War elicited significant support drawing upon a wide range of different groups, some of them religious-based others more politically orientated. Such unexpected coalitions are themselves of interest and support the view that traditional boundaries are becoming blurred as former enemies make common cause on issues of violence and injustice.

What is less clear is the impact that such movements will have and whether these temporary coalitions will endure and be able to effect significant political change. One might also question whether such influence is either desirable or justifiable given the possibility that those who participate may simply be the vocal and articulate who are not representative of the wider populace. The capacity to exercise power does not, in itself, justify actions taken or influence gained. There is a tendency in much of the recent literature on global civil society to see it as automatically a 'good thing' and to present it as if it were more coherent and organized than it actually is. Keane reminds us that there can be bad voluntary organizations just as there can be good companies and governments. Jumping on bandwagons is the phrase that comes to mind here. Critics of those who now advocate environmental governance on a global scale fearing the effects of rapid climate change and global warming are quick to point out that the scientific evidence on both sides is presented in such a way as to support an already adopted position and that greater ambiguity and uncertainty is an inescapable feature of scientific research. NGOs can easily become power blocks of their own with a vested interest in continuing 'scare stories' and stirring up public opinion without rigorous reference to the actual arguments and evidence.

Is there an emerging global civil society then? Much depends on the definition of terms as ever, but there are certainly significant trends of human association across national boundaries that are neither government nor business driven and might suggest that something new is happening. However, the detail on this is somewhat sketchy so far and one must refrain from making exaggerated claims for these movements. Engineers and scientists as individuals and also sometimes as part of groups with their own transnational networks and concerns for peace, justice and the environment are to be found within these movements but often not at the head of them. Global advocacy is itself a challenging role to play and requires attention to detail in ways that are not familiar very often to those of moral commitment and who are sometimes keener to show their involvement in worthy issues than to examine the evidence. Critical questions must be part of what such groups bring to these issues even when they lack the relevant expertise. Asking ourselves where we should stand on these concerns and why is surely a vital task and it is worth reflecting on how good we are at doing this.

3.11 THE LOCAL AND THE GLOBAL

It is possible to argue that globalization pulls most of us in conflicting and potentially contradictory directions, both outwards towards the wider world and inwards and local where matters seem more proportionate and manageable. The American sociologist Rosenau provides some important insights and analytical tools on this issue (Rosenau, 2003). He argues that the process is one of both integration and disintegration at the same time. What seems remote can also now seem close at hand, hence his term 'distant proximities' which he prefers to 'globalization' as a description of where we now are. We need to see the world as an endless series of distant proximities in which the forces pressing for greater globalization and those inducing greater localization play off against each other through constant interaction. So there are tensions between core and periphery, national and transnational, communitarian and cosmopolitan, cultures and subcultures, urban and rural, decentralization and centralization, the universal and the particular and so on (Rosenau, 2003, 4–5). We are caught up in the midst of diverse ways in which societies respond to new developments in economy and culture. It is not that global is good and local bad, or vice versa, because no such general evaluation is possible. Rosenau coins the term 'fragmegration' to describe these complex and contradictory forces. A move towards the local will not necessarily mean greater fragmentation just as a move towards the global will not necessarily mean greater integration. All of this is similar to the ideas of Beck and Castells, but what Rosenau does also provide is an interesting categorization of how different people then respond to these changes.

Given that the dynamics of globalization are boundary blurring and complex, it is more likely that the dynamics of localism are going to be boundary preserving and that those who are keen to operate at a more local level are orientated towards proximate horizons rather than distant proximities (Rosenau, 2003, 81). Identity formation may still be tied to a particular place or locality, partly in resistance to global forces. Thus there are different types of locals. Insular locals are much less affected by globalization, strongly based in a particular neighbourhood or territory, perhaps centred around family or local church and closely identify with a community whose boundaries are fairly clear and secure. Whether there are many people who now still fit this description is perhaps doubtful, but the category itself still operates in our view of the importance of locality. There are also however affirmative locals, those who have come to terms with the processes of globalization and probably move around the world either physically or virtually with confidence and ease and yet still retain a sense of the local and operate at these levels.

Further categories are resistant locals and exclusionary locals, both of these eager to resist the effects of global change, perhaps fighting to prevent jobs being lost overseas or to retain the specifics of a particular culture. Issues such as immigration and national sovereignty might figure high upon their political agendas. So the local is seen as the battleground on which to contest the erosion of values and institutions. Anti-globalization protests may fit somewhere within this, although of course the protestors themselves use global travel and global communications in order to make their point. Exclusionary locals are likely to search for their identity through ethnicity, nationality, language and culture or indeed religion—hence perhaps the rise of 'furious religion'. Rosenau also provides different categories of 'globals' but those are perhaps of less relevance to this particular debate.

Where might those of moral commitment fit on this spectrum of 'locals'? Might the retreat to the local perhaps represent simply a resistance to what is really happening in our lives and carry with it an unreality and illusionary security bound up in what appear to be traditional terms and practices? Yet is not some form of local base and identity a prerequisite for effective and confident engagement with global change? How can we know the difference between the two? Are we 'locals' for the sake of being local or rather locally based as a means of taking insights onto a global stage? There are perhaps no straightforward answers to such questions, but they are nevertheless important ones as we try to work out what are appropriate responses to fragmegration and our engagement with distant proximities. To what extent are our 'identities' changing even though we may not be prepared to acknowledge it? What is the future for institutional articulations of morality when so much now is thrown back on individual decision? Have we yet fully grasped the implications of 'globalization' and its impact upon engineering identity and practice? Each of these questions requires our critical attention.

3.12 GLOBALIZATION AND RESISTANCE

In this final section on globalization we return briefly to the question of whether what we are really talking about is an American takeover of international power through economic means. For a critical response to this I turn to the work of two post-Marxist writers, Hardt and Negri, who challenge this view (Hardt and Negri, 2000, 2004). The background to this is formed by a series of publications following the aftermath of the collapse of communism and the end of the Cold War. The argument was that, with the demise of communism, the only player left on the world stage was Western Global Capitalism in the form of various democracies, especially the USA and the developing EU. Thus there was no longer any serious political alternative and what follows is bound to be the further development of democracy combined with capitalism. Hardt and Negri have a different interpretation of events and use the term 'Empire' to describe this, but they are not referring to a particular nation, i.e. the USA when they do this. Rather they believe that no one nation state, however powerful, can achieve that sort of global dominance and control given the complex dynamics of globalization. The deregulation of labour and capital that has been required in order for capitalism to sustain itself over the last two decades has in fact undermined the capacity of the nation state to control the flows that now cross all boundaries and state borders. This does not spell the end of sovereignty however, but merely its reconfiguration into this form called 'Empire' in which there is no longer a clear 'inside' or 'outside' to the new world system. Empire has no centre and no limits, but is a decentred and deterritorialized apparatus of rule that progressively incorporates the whole globe.

The basis for this is a three-fold level of organization that mirrors Polybius's model for the operation of the Roman Empire. There were then three principal forms of government: monarchy, aristocracy and democracy. Their modern day successors are represented by the IMF, World Bank, WTO and NATO as monarchy: transnational capital as aristocracy and nation states and NGOs as democracy. Capital now subsumes the whole of social life as a result and the emerging juridical constitution oversees this process, supervising and regulating the relationships at a global level. In some ways this appears to be a benign process, welcoming all into the fold, regardless of race, colour, creed or gender, but the effect of this is to stifle opposition and prevent effective alternatives. So the operational principle here is no longer 'divide and conquer' but rather 'absorb and manage', all within the ever-expanding clutches of western capitalism.

Although this might appear to be a very depressing and deterministic scenario, Hardt and Negri are convinced that this means of power can and will be resisted no longer by a particular class or group of people as in the communist master plan, but by a range and diversity of sites of resistance themselves characterized by complexity and the blurred boundaries that now dominate globalization generally. Holes will be punched in the system in myriad different ways and places on the basis of quite different protests and alliances from before. There are questions about the substance of this part of their argument, but there is a logic to it nevertheless.

Do these ideas provide any sort of hope for the future or a viable alternative for those who would oppose many of the worst excesses and injustices of the global system? It is possible to employ Hardt and Negri's ideas about 'enclosures' and 'thresholds' in a value-based analysis of globalization I would suggest. We might ask ourselves where and why people are now trapped within this system and even unable to raise critical voices and questions about its hegemony. The answers to this will reveal the areas of injustice and economic dominance of political powers but also ways in which our ideas and understandings are themselves in thrall to capitalism and its gods of consumer choice and short-term profit.

The bigger issue is that of where we go from here and whether there are realistic prospects of either overturning global capitalism in some coherent way or either 'humanising' it in ways that protect both human well-being and that of the planet which we share with other living creatures. Will there be incremental or revolutionary changes or will the system self-destruct in any case as we reach the environmental limits of capitalist growth? What should people of integrity make of all this and how do we respond given the complexities of our own communities and beliefs? To help address some of these issues we need to turn to a further set of resources and to consider where engineers might more effectively locate themselves within these changes that are encompassed within the arguments about globalization.

CHAPTER 4
Prospects for Creativity

4.1 DIRECTIONS FOR INNOVATION

As we begin to draw together the threads of the discussion, I want to offer another example of what I believe to be the future direction for engineering. The Ford Motor Company has made the decision to 'go green' on a serious scale. During the summer of 2006 its UK branch announced that it was going to invest £1 billion in research and development of a new model that will reduce the amount of petrol required and thus respond to environmental concerns. In fact, if one looks at the official Ford website this is only one of a series of initiatives that are now underway worldwide.

> At Ford, we take concerns about climate change and constraints on the use and availability of carbon-based fuels very seriously. Since 2000, we have cut the emissions of CO_2 from our plants and facilities by 15% and we have targeted further reductions.

In addition, and perhaps more significantly, the company is in the process of developing new models which include hybrid vehicles, cleaner diesel versions and a hydrogen-based model. While this is all very worthy and it is clear that companies now have to be seen to be attempting to become environmentally friendly because of government policies, there is also something else driving this process—global competition. The reality is that Ford is starting to lose out to rivals that are already ahead of the game in the green stakes. Toyota pushed Ford sales to third place for the first time in the summer of 2006, largely because it has forged ahead in terms of developing and putting into the market vehicles that meet environmental concerns and have an advantage given the recent rises in fuel costs. So, on one level, it is simply global market forces and changing consumer habits that force businesses such as Ford and Toyota to innovate through the development of green technologies.

It would be unwise to dwell too much upon the claims and counter-claims of particular companies, but, that such global giants of the motor industry are now moving in this direction suggests that environmental concerns are going to be central to future innovation. In a global economy, even the New Economy which creates its own restrictions for the profession, there are still external forces which will shape the future of innovation in a positive manner. It may all be a case of too little, too late of course, but I would argue that it is reasonably clear that environmental problems produce the major factor now shaping creativity within the profession. Is there more that can be said about this? I think that there is and I will aim to articulate exactly what within this chapter.

4.2 ENGINEERING AS A TRADITION

Traditions have to change in order to survive. If that seems an obvious statement, it is perhaps less obvious that it should be applied to the varied skills and frameworks that go to make up the engineering profession. After all we are surely talking about the application of ideas and skills rather than some form of intellectual progression which we might more readily identify with the sciences. Well, perhaps, but I would argue that engineering does possess the characteristics of being a tradition. It has its own identity over time based on certain principles and interpretations of how the world works: it passes on those ideas through its own processes of education; it certainly claims to embody sets of values and ideals about how it serves the wider community and it 'polices' its own boundaries through constant debate about what falls within its remit and how changes are to be negotiated. Is it aware of itself as a tradition though? Again, I would say that its current capacity to reflect upon the context within which it is operating and to develop appropriate ways of responding suggests that it is. The very fact that there is a text such as this, which is itself part of a series examining the wider questions now facing engineering, is a consequence of engineering being a tradition that is capable of looking critically at itself and examining some of its presuppositions and the external factors which shape its progress.

If this is correct, then one of the most important impacts of globalization upon the profession is that it is moving it towards being what social scientists would call a 'reflexive tradition'. What is meant by this and how does this differ from what has gone before? In order to answer this question I will introduce some ideas from philosophy which will illuminate the discussion.

A UK-based philosopher, Simon Critchley (Critchley, 1999, 136), has helpfully provided a typology of approaches to tradition. His aim was to examine the tradition of philosophy itself, but I believe one can argue that his ideas have a wider application. He talks about sedimented tradition, reactivated tradition and then deconstructed tradition—somewhat technical terms that I will describe in more straightforward language. Sedimented tradition refers to the 'forgetfulness of origins' that characterizes a stage before which a tradition is even aware of itself as such. So it is simply taken for granted that 'things are as they are and always have been'. If questioned, its adherents and advocates are likely to respond along the lines of 'well, this is the way we have always done things here'. Even the values which may be part of the tradition are not normally brought to the surface and examined in any conscious or deliberate fashion. In another context I have referred to this as 'practical consciousness' (Reader, 1994, 20). This is just 'normal life' if you like, the routine pursuit of a way of doing things that is not particularly bothered about or even aware of its own history, processes of development, or ways in which it might be subject to challenge and change.

When writers such as Beck, who was mentioned in the previous chapter, talk about the need to become more reflexive, part of what this means is that it is no longer possible, let alone desirable, that traditions should continue in this almost semi-conscious manner. There is more to it than this as we shall see shortly, but this is the beginning of a process of creating greater awareness of the way traditions operate. That Ford Motor Company can even examine the world in which it is set and make the decision to 'go green', even if that is a response to market forces, makes it clear that engineering has not remained at the level of a sedimented tradition. The tendency to shut one's eyes to the wider operating environment would simply be a recipe for bankruptcy as far as any commercial business is concerned. There is no choice but to be reflexive in that sense.

So what about reactivated tradition? This is the point where a tradition becomes aware of itself as such and begins to question what legitimately rests within its territory, frequently when under challenge from external forces. The exercise of human autonomy, powers of reasoning and critical questioning each fit within this description of tradition and quite obviously apply now to the engineering profession. In fact, one might argue that these have always been essential to its development at a technical level. However, this should now be extended to the dimension of values and principles perhaps embodied in codes of conduct and standards. What is this profession really about, whose interests should it serve and what are the influences that challenge its identity and integrity? These are the sort of questions which characterize a reactivated tradition, one that is prepared to take its presuppositions 'out of the box' to examine and question them and then to reflect upon its current and future directions. The decision to 'go green' could simply be a response to market forces, or it could also be a matter of principle. Either way, it reflects a tradition conscious of its own location in the world and open to new possibilities. One might say that creativity and innovation can only occur within a reactivated tradition.

This all sounds reasonably straightforward and familiar I would hope. However, there is a further dimension which takes us beyond this critical self-awareness. Although I do not intend to go into the technical definitions of 'deconstruction', which, to the uninitiated, always sounds as though it actually means destruction, it will be necessary to refer back to this later in the chapter. I will use it there to talk about a personal level of self-awareness and development that goes beyond the sort of institutional and collective dimension that we are referring to now. Suffice it to say that it is another way of responding to the challenges of a world which presents us with complexity and confusion, that blurring of boundaries that Beck mentions in respect of reflexivity. One might say that engineering at an institutional level will start to wonder whether it is not quite as it imagines itself to be, whether its values and modes of operation actually contain other elements and influences that contradict or work against its perceived identity and even undermine its integrity and purpose. In which case it will be forced to look again at how it responds to external forces in order to become aware of the 'internal other' that affects its image and operation.

Returning to the language of enclosures and thresholds, one might say that if engineering were to remain a sedimented tradition that would make it an enclosure that restrains and restricts future development and progress. Although one might assume that moving into being a reactivated tradition automatically means encountering thresholds and new possibilities, the reality is more complex than that. One of the arguments of this text has been that the encounter with globalization in the form of the New Economy (globalism perhaps) is in fact more of a new enclosure than a threshold, bringing its own new and unforeseen restrictions that the profession needs to be aware of if it is to protect its core values and codes of practice. Again, one is forced to the question 'whose interests are being served here?' in the economic system as it now operates. Is the answer consistent with the self-understanding and identity of a profession which claims that it is there to serve the well-being of others? Perhaps a deconstructed tradition is one that is better equipped to ask such questions and to consider its place in the world. To take this discussion further though it is time to look more closely at the theme of reflexivity.

4.2.1 Reflexivity and Sustainability

It might be tempting to think that environmental issues are no more than the latest fashion or, at best, a significant challenge that can be met by engineers in terms of research and development. However, there is a growing awareness that what we are now facing is, in fact, a more profound question mark about the direction of our global culture. A term which encompasses those concerns, at least in the UK, is 'sustainability'. So far we have tended to suggest that globalization in the form of globalism creates as many problems as it solves in terms of justice and morality, let alone working conditions for engineers and others. However, there is still the belief that other forms of globalization and a revised economic order could deliver stability, security and the continuation of the current social system in a recognizable form—hence the comments of advisors such as Stiglitz. That position is becoming more difficult to support as we become more aware of the levels of challenge now facing the human race on this planet. Put briefly, in addition to the projected problems created by climate change and global warming, it is now clear that we are entering a period where it is uncertain to say the least, that we have access to the resources that are required to maintain 'business as usual'. Questions about the timing of peak oil, locations of oil and gas fields and who controls those are clearly central. Whether the current growth levels of China and India are sustainable, either economically or politically, is another major factor in the agenda of uncertainty. Both of those are also linked to the general threat of military activity either of a terrorist nature or the responses of national states and the role of faith groups who pursue their objectives with an increasingly apocalyptic zeal.

Into this scenario ride the various 'saviours' offering hope of not so much a better future of greater overall wealth and prosperity, but technical solutions that might just pre-empt the more disturbing political prospects of conflicts over access to the remaining natural resources that literally 'fuel' the global economy. Given that so much of the world that is taken for granted, particularly in the developed nations, rests upon the innovations and skills of the engineers of the last 200 years and their impact upon physical infrastructure, medical technology, transport and global communications, it is hoped that further innovation will now prevent a descent into political conflict that would threaten the advances of industrialization itself. Reflexivity then is something rather more than a luxury recommended by intellectuals who wish to stand back and review the situation for its own sake, it is about survival and sustainability.

Needless to say, as is common with any academic discussion, the term is open to a number of interpretations and the usual mix of arguments and conflicts. For the sake of clarity and brevity I will steer a particular path through these and present only the aspects that I believe are most valuable for this discussion. An example of what some sociologists mean by the term would be the practice of re-cycling information now readily available in order to facilitate the making of personal or collective decisions. So, for instance, those considering getting married would first study the statistics on marriage breakdown before making their decision; those who are diagnosed with certain conditions or are concerned that such conditions run in their families and wish to know what the chances are of them inheriting the condition, will study the information available on the Internet. It is simply a matter of making the best use of the vast amounts of information now available to us so that we can make informed and rational decisions. However, other writers are less optimistic about this process and draw attention to the hidden dangers that we now face.

Beck in particular draws attention to the unconscious dimension of reflexivity:

> The reflexivity of modernity and modernization in my sense does not mean reflection on modernity, self-relatedness, the self-referentiality of modernity, nor does it mean the self-justification or self-criticism of modernity in the sense of classical sociology; rather (first of all) modernization undercuts modernity, unintended and unseen, and therefore also reflection-free, with the force of autonomized modernization. (Beck et al., 1994, 176)

So this is not a matter of reflecting upon the social conditions of our existence in order to change them, but of the processes of change that are going on 'behind our backs' through the unintended and unforeseen consequences of our actions. Hence it is not the same as an optimistic or cognitive version of reflexivity that would argue that more reflection, more experts, more science and technology and of course more self-awareness and self-criticism will deliver a better world in line with our expectations. This is a much more challenging and disturbing view of reflexivity but also, I would argue, more realistic. On one level, we have no choice but to be reflexive now as we have already seen, but that in itself is no guarantee that the decisions we make in response to that greater awareness will lead to the consequences we desire or predict. We are entering essentially contested territory where issues of power will confuse the outcomes and where different groups may well be in conflict over exact interpretations.

4.2.2 Reflexivity and Autonomy

> The talk of 'unintended consequences' thus denotes a conflict of knowledge, a conflict of rationality. The claims of different expert groups collide with one another, as well as with the claims of ordinary knowledge and of the knowledge of social movements. The latter may well have been developed by experts, but according to the hierarchy of social credibility, it is not considered to be expert knowledge and, consequently is not perceived and valued as such in the key institutions of law, business and politics. The knowledge of side-effects thus opens up a battle-ground of pluralistic rationality claims. (Beck, 2000, 120)

Perhaps this takes us beyond our current brief, but it is surely a salutary warning that the application of certain technologies cannot be relied upon to deliver agreed objectives or expectations in a linear fashion. Providing an apparent solution to the proliferation of greenhouse gasses by developing more environmentally friendly vehicles cannot be guaranteed to provide the answer to the problems of climate change, which is not to say that this route should not be followed, but that we should not be naïve enough to imagine that this may not lead to other even less palatable consequences, certainly when they are set in the wider context of global capitalism in its current form.

Essentially this is a debate about the role of autonomy and reason within our culture and our understanding of what it is to be a human being. If this sounds a long way from our starting point then so be it, but it is a dimension of the debate that becomes less avoidable once one takes seriously the challenge of globalization. A reactivated tradition, capable of standing back from its 'traditional' self-understanding and practices, which is what I am suggesting engineering is now becoming, might imagine that it can provide a 'technical fix' for the problems we face. A bit more analysis, combined with a heightened awareness of the questions of values and codes of practice, is what is required if our current cultures and lifestyles are to be 'sustained' and issues of the environment, poverty and justice are to be addressed. Is this not though to place too great an emphasis and reliance upon the powers of human reason and the supposed autonomy which places humans at the cutting edge of the evolutionary process? Why should engineers worry themselves about such matters at all?

Returning to the environmental debate, if one digs a little way into its recent history, it rapidly becomes clear that one of its roots is as an anti-science and technology movement. It is precisely the exercise of human reason and autonomy through the application of scientific research and its handmaids of engineering and technology that lies at the heart of the ecological crisis. What some environmentalists have argued for is a deeper understanding of what it is to be a human being that either returns to a simpler and earlier (romantic) view of a closer relationship with the natural world, or moves beyond modern culture to a further stage of human development. These strands of thought are still present, even though science and technology have now largely appropriated the environmental territory or been brought in as allies in finding solutions. But there is an internal contradiction here, or the presence of an 'internal other' in this debate that represents exactly the sort of 'deconstructed tradition' that Critchley suggests goes beyond the reactivated versions. A genuinely reflexive tradition would need to pay attention to this even if it could not see any obvious resolution to these differences.

In the final part of the text I will offer a framework that will help clarify this position and provide a means of moving this discussion forward. But before I do that, it would be useful to examine some other recent research on the subject of human well-being that brings the debate to a similar point but from a different angle.

4.3 THE DEBATE ABOUT HUMAN WELL-BEING

The question of what it is to be a human being and ways in which we might answer that from within our culture do seem somewhat distant from our original concerns. However, if we recall that one of the issues to emerge from the early discussion of the current state of engineering was that of the increasingly restrictive working environment for engineers and subsequent problems of recruitment and retention, let alone that of motivations and values, then it is difficult to separate that from the question of human identity. How important is job satisfaction in people's lives? How much of a role does one's working life play in overall measurements of human well-being? How significant are the financial rewards that the New Economy and current versions of globalization present as being the main objective of business and commerce? It is these sorts of areas that are now being examined by economists and are taking a high profile in discussions about development economics in particular.

Work carried out by the UK economist Layard has revealed some interesting findings. Since the 1970s in both the UK and the USA, figures indicate that there has been almost no change in happiness despite a doubling of living standards. His conclusion is that, above a certain level of income, any increase in prosperity is unlikely to generate an equivalent increase in happiness.

> Our study suggests that if everyone else earns another 1%, your happiness falls by one third as much as it would rise if you yourself earned an extra 1%. So if everybody's income rose in step, your happiness would rise, but only two thirds as much as it would if only your income was rising. Another study probes the issue more deeply, by examining what really affects your happiness: is it your actual income that matters, or how you perceive it to compare with other peoples? Your 'perceived relative income' shows up as more important than your actual income. (Layard, 2006, 46)

So this psychological mechanism reduces the power of economic growth to increase happiness and results in distorted incentives. If I work harder and raise my income I make other people less happy. Yet this is not taken into consideration when most of us decide how much we will work and the tendency then is to work more than is socially efficient. The actual figure quoted beyond which this marginal increase in happiness mechanism starts to operate is between $15,000 and 18,000. Obviously there are a whole series of variables and complexities which will lead to differences between individuals and between professions, but the general point still holds that simply increasing income beyond a certain level does not guarantee any greater sense of human well-being. In which case one might begin to question the driver of profit which apparently controls the global economy. Why is there this addiction to increasing income and profit if, in fact, it shows little evidence of making people any happier?

This shifts the debate to a further level and raises the question of what factors do influence and determine human well-being. Anybody familiar with Maslow's 'Hierarchy of Needs' will recognize that much subsequent research develops different and expanded versions of his original contribution. For those not acquainted with this, his research revealed that there are a number of levels of human need and once one has been satisfied, and only then, do other needs take on a priority. Thus basic levels of need revolve around physiological requirements of food, shelter and nurture, followed by the need for security and safety and then a sense of belonging and community. Higher levels then move onto the planes of aesthetics, learning and discovery, esteem and self-respect, self-actualization and a further level which is that of transcendence with its spiritual and religious connotations. This is important for our discussion about creativity within the engineering profession I would argue. The sort of satisfaction gained from being involved in creative and innovative work is clearly a higher level need and one that can only figure once lower level needs are satisfied. Thus if your immediate priority is physical survival either because you do not know where the next food is coming from or where the next missile is likely to fall, you are less likely to be concerned about exercising your creative faculties. Yet if the objective of our culture is to ensure that basic needs are met, then it is highly likely that people will start to require things like job satisfaction and creative opportunities and feel demotivated and frustrated when these are not acknowledged. Reducing the rewards for work to simply financial factors is going to be of limited value in more developed societies.

Just to give a sense of the sort of factors that have been identified in research projects since Maslow and that are now being used within development economics as ways of evaluating projects in developing countries I refer briefly to the work of Sen as documented recently by Alkire (2005, 78–84). Most of the typologies presented in this text build upon Maslow's basic structure, but with different emphases and expanded areas of interest. It is probably the areas of human relationships, of creative expression, awareness of a closer relationship with the natural world, opportunities for autonomy and independent judgement, and then personal and spiritual development that add most to the original ideas. Challenging work is also mentioned as one might expect. All of this adds strength to the argument that there is more to life than simply money or power and that human beings require satisfaction and stimulation at much deeper levels if they are to realize the potential that lies within. If economists, of all people, now take such ideas seriously then one can only hope that the realms of business and commerce, let alone science and technology, begin to recognize the importance of studying the subject of human well-being and identity.

What I intend to do finally is to link this research back to the earlier discussion about the potential difference between understandings of the science, engineering and technology and environmental issues and views of human nature which underpin them and offer a framework which will provide a stimulus for ongoing debate.

4.4 THRESHOLDS TO CREATIVITY

Are there then aspects of globalization, understood in its wider sense and thus not restricted to the globalism associated with the New Economy, that might encourage or enable creativity within the engineering profession? If there are then I propose that they are most likely to emerge through the route of a deconstructed approach to the tradition, one which is prepared to acknowledge the tensions and internal conflicts which are the result of operating in a commercially dominated environment and to address issues of the environment linked to questions of human identity. To put this in more familiar terms, we are talking about the area of personal development, placing this within the broader notions of transcendence or spirituality brought out through recent studies of human well-being. None of this is to say that innovation and creativity will not be possible without those elements but rather to suggest that the blurring of boundaries characteristic of globalization will push creative engineering in this general direction.

A definition of creativity is worth offering before we progress. This comes from the writing of Roy Bhaskar, a philosopher of science who has been central to the movement known as dialectical critical realism, but who has recently moved into more esoteric areas of research utilizing eastern religious traditions (Bhaskar, 2002, 122).

Bhaskar promotes the idea of the scientist as a practical mystic and describes creativity in the following manner:

> The creative scientist becomes a kind of mystic, a practical (scientific) mystic. Moreover it shows that science necessitates intuition and the intuitive alongside the discursive intellect, iconic or pictorial, imaginative and visualizing activity as well as sentential formation and discursive conscious activity. For alongside left brain (deductive/formal—instrumental/calculative/manipulative thought—deeply embedded in structures of oppression... discursive intellect), there is another 'right brain' intuitive power of discrimination—the intuitive intellect. Both left and right brains are necessary. (Bhaskar, 2002)

I quote this because despite its complexity it highlights the problem faced by all attempts to critique the reactivated tradition based on ideas of human autonomy and powers of reasoning. If engineering is not to retreat to some form of romantic pre-autonomous level of human functioning—and that would seem impossible anyway, but it is the sort of move advocated by some environmentalists in their opposition to science—then it needs to both acknowledge the importance of reason and autonomy and recognize that they are not sufficient to respond to the global problems we have created. Only a framework which includes all levels of human being, something similar to Maslow's, but allows us to recognize the contribution of each while still encouraging individuals to press beyond each level, is adequate for the task. Bhaskar talks about the importance of both left and right brain, the discursive and the intuitive intellect. I will expand this now by drawing attention to the work of Ken Wilber who pursues a similar approach but building upon an even wider range of sources.

Having researched all the major spiritual traditions still extant and attempted to produce his own typology for spiritual or personal development, Wilber proposes that there are (at least) six basic levels of human consciousness, and these are in ascending order of experience:

- The material level, at which consciousness is identified exclusively with sense data, and the self and the physical body are seen as synonymous—failure to progress beyond this level means there will be no awareness of any underlying unity.
- The vital level: this is when consciousness becomes aware of the self and the distinction between life and death. So life is valued and death seen as either a threat or a mystery.
- The discriminative level: so the consciousness begins to categorize objects and events encountered in experience and to recognize differences between the purely physical and the world of inner experience, thoughts and perhaps spirituality.

- The ratiocinative level: at this point the self develops the capacity for analytical and rational thought, becomes capable of abstract thinking and advanced theory building.
- The causal level: consciousness now experiences pure contentless awareness, so rather than being aware of something, consciousness is simply conscious of itself—this is to access higher levels of awareness normally associated with meditation and spiritual practices.
- The Brahmanic level: consciousness is aware of reality as a united field of energy in which the material world, the individual and the source of all life are in essence identical with each other (Fontana, 2003, 177).

In other words, as far as science and engineering are concerned, one can recognize the level at which both operate, Level 4 where the self develops the capacity for abstract thought and advanced theory building. This is fine, as far as it goes, but it can fall into a form of reductionism if it fails to move beyond this, and is always prone to being appropriated by other motives if it is unable to see forward to further levels of human development. Creativity at this level is clearly of potentially great benefit to economic and social development and has contributed significantly to the culture that we now take for granted. Yet it can nevertheless get stuck at that point, become subject to the purely financial considerations under the influence of the New Economy, and then cease to provide the satisfaction that was probably present in earlier eras of industrialization. So the challenge is to take the profession into new and relatively unexplored areas of human development, probably initially through a concern for environmental issues, but aware that those also are only one level further on and can even represent a regression if they fail to acknowledge the role of human autonomy (Bhaskar's left brain discursive intellect).

Whether the engineering profession or indeed any of us are ready for this and able to make connections with where we are now is of course an open question. But then this is a threshold and we have the choice of either passing through it or remaining where we are now. The latter option leaves questions of justice, environmental concern and sustainability unanswered and the future very much in the balance.

I hope that this swift journey through the complexities of globalization and related issues has provided enough evidence for engineers and perhaps scientists to think carefully and critically about the context in which we are now operating. Without making the mistake of adopting some form of economic determinism which states that the global economy is bound to move in a particular direction, one might begin to see that particular forms of globalization are detrimental to the engineering profession both in terms of working practices and prospects for innovation and creativity. Not only that, the wider contribution to the well-being of others which is surely a core value of the profession is extremely difficult to protect in a context where short-term profit appears to be the prime directive. Yet there are always thresholds, openings to new possibilities, even within the enclosures that constitute any political and economic system. The challenge is to identify those and then to find ways of working towards them. Developing the self-understanding of the profession itself is an essential response to this challenge, but then so is the individual level of a deepening self-awareness and the move towards a more reflexive identity. The capacity to keep thinking and to keep feeling is central to this task.

Glossary

Civil Society: Used by Political Sociology to refer to the level of human organization that fits between the state and the family. So business, voluntary groups, non-governmental organization and professions would all fall into this category. There has been a renewed focus on the importance of Civil Society since the fall of communism in 1989.

Cosmopolitanization: Ulrich Beck's term to describe the dimensions of globalization that go beyond the purely economic: the blurring of boundaries which comes from increased interaction between different nations and social groupings as a result of ease of travel and communications and creates hybridity and the 'salad bowl' mixing of different lifestyles, beliefs and cultures.

Enclosures: Taken alongside the term 'threshold' as adopted from the ideas of Hardt and Negri, it refers to the problem that the new economic order encompasses all aspects of human social and economic life so that there is no longer any 'outside' to the current system. Even what present themselves as areas of opposition and resistance to global capitalism are in fact already within the system according to this theory. 'Empire' is the enclosure in which we are all now set and thus produces constraints that we may not even be aware of.

Enlightenment: A term used by philosophers and historians to refer to the movement of thought which began in Europe in the eighteenth century based on a view of human reason and autonomy which challenged the authority of religious traditions in particular to establish understandings of truth. So science, engineering and technology could be said to derive their independence and authority from the revolution in human thought begun by the Enlightenment. It is also often associated with a belief in a linear process of human progress achieved through rational thought and political order.

Fragmegration: Rosenau's way of describing the double movement of both integration and fragmentation, which, he argues, characterizes what we describe as globalization. So one cannot argue that greater globalization automatically creates fragmentation nor that localization creates greater integration. Both are identifiable results of the tension between global and local that we experience at all levels of society. A useful term because it highlights the complexity of the processes that we are trying to understand and analyse.

Globalization: Held's definition is probably the most straightforward: "at its simplest refers to a shift or transformation in the scale of human organization that links distant communities and expands the reach of power across the world's regions" (Held, 2004, 1). However, as we have seen in this text, it tends to be identified with a particular form of economic development or set of policies and thus becomes a highly contested and controversial term. Perhaps the best advice is to 'handle with care' and make sure one is clear about the aspects of globalization to which one is referring.

Globalism: In reaction to the above preferred by some authors (Stiglitz, Beck) to refer directly to the growth of global economic capitalism associated with the New Economy. Thus it is based upon short-term profit and shareholder value as the key measures of economic growth and often leads to the downsizing of labour forces and the enforcement of a more rigid employment regime under the threat of shifting production to parts of the world with lower labour costs. Potentially a threat to creativity and innovation within the engineering profession as industry shuns the risks involved in longer-term investment

Governance: A term that is becoming increasingly popular within the globalization debate to refer to the structures of control and monitoring that might be able to contain the worst effects of economic growth. So it is encountered in the discourse of environmental debate and also that of the movement of capital around the world and the conduct of big business. Whether or not governance can be an effective counter to corruption and human error is a matter of argument, but sociologists such as Held set great store by it as a rational response to global problems.

Ideal Types: This description was coined by sociologists (Weber etc.) to refer to ideas that never find a direct equivalent in the real world. They are important as they help us to categorize and then discuss social action and social movements, looking at them from a distance and drawing out general trends that do not necessarily appear in any single instance. However, it must be remembered that they are at one remove from reality and are limited tools of interpretation only.

Instrumental Reason: Another technical term from within the field of philosophy which represents the view that reason as it has been shaped in recent decades is of a purely 'means-ends' nature. So once an objective is agreed, the only role of reason is to work out the best and most efficient means of achieving this. So issues of values or morality do not enter into the realm of reason at all. We have seen in the text how public perceptions of science and engineering are damaged to the extent that ethical considerations are not taken into account. In other places I have argued that there are wider understandings of reason now available that offer a better basis for moral action (Reader, 2005).

Internal Other: This is an unfamiliar phrase to engineers that comes from the worlds of philosophy and psychology. It refers to the realization that each individual contains personality traits or bases for behaviour that appear to contradict or conflict with their normal external appearance or self-understanding. It can also be used to talk about institutions or other forms of social structure or bodies of belief in the sense that it is always possible to identify strands that are not consistent with the dominant identity of each. The internal other can both be a cause for pessimism in that it suggests there is little one can do to counter such contradictions and a source of optimism in that it suggests that there is always the hope of other possibilities breaking through (thresholds). The current internal other of engineering could be the forces of globalism that conflict with its values and codes of practice but yet might create a counter-reaction that leads to greater awareness and resistance.

Labour Flexibility: Linked to the arguments about the New Economy are recommendations that any work force must accept the need to remain competitive in the global economy. Hence there is a need for flexibility both in terms of skills, time frames for projects and employment contracts and a willingness to forego the protection that Trade Unions might have offered under a more favourable regime. Thinkers such as Sennett argue that this flexibility is actually damaging to 'good work' and to the individuals involved, hence his comments referred to in the text about craftsmanship.

Liquid Modernity: Zygmunt Bauman's work on the sociology of modernity has established this phrase within the discourse as a way of describing the speed and fluidity with which life now moves. Liquid is to be contrasted with solid in terms of relationships, institutions, working practices and even culture. In global modernity everything is 'for the time being only' and 'until further notice'. As with Sennett it leads to the insight that there is a deliberate instability which serves the economic interests of some at the expense of others. Although a powerful idea one needs to identify areas where complexity is greater and there is still effective resistance to this rapid movement.

Meta-narrative: A meta-narrative is a proposed universal framework of explanation or world view which claims to encompass the most important aspects of human knowledge. So many world religions have certainly fallen into this category but so would a scientific approach which made a claim to provide a 'theory of everything' or to offer a definitive understanding of human existence or pathway to truth or objectivity. Post-modernity in its philosophical mode is the argument that we are now in the time of the breakdown of meta-narratives where all such claims to truth can often be interpreted as bids for power. In an extreme form it leads to relativism, but other options are possible.

New Economy: The term invariably used by economists and politicians to refer to the increased influence of information technology and the importance of networks etc. However, it is also frequently identified with the neo-liberal, de-regulation agenda of the IMF, World Bank and World Trade Organization as each of those falls under the sway of U.S. economic policy and is therefore seen as damaging to other nations within the global system. The question posed by Castells and Stiglitz is whether this is the only or inevitable shape for the global economy or whether different decisions might lead to a different form of 'New Economy'.

Reflexivity: An idea which now covers a range of possibilities but which, at its simplest, refers to the human capacity to use readily available information to help make important life decisions—e.g. accessing sites on the Internet in order to discover more about a particular medical condition, chances of survival, etc. I have used it here with elements drawn from the work of Beck who points to the limitations of a reflexivity that is no more than a conscious reflection. Reflexivity also means the unintended consequences of our actions and the things always out of control that impact upon our best laid plans. So although globalization increases our self-awareness and our capacity to think critically about our world, this reflexivity does not lead to instant or clear solutions in a linear or reliable fashion. It is also part of the continued blurring and complexity of modern life.

Thresholds: If we take this term in combination with 'enclosures' we can see that there are both constraints and new possibilities open to us as we encounter the forces of globalization. Where precisely the thresholds will appear is the challenge that now faces those of us who wish to resist the growth of global capitalism in its current form. Without therefore going once again into detail I would simply add that if one does not believe that there are such thresholds then one is acquiescing to a form of economic or political determinism—i.e. accepting the view that 'there is no alternative'. I would have thought that engineers thrive on the challenge of finding alternative solutions and other ways through the labyrinth.

Tradition: I am aware that I have used a complex interpretation of this following the ideas of Simon Critchley, hence his notions of sedimented, reactivated and deconstructed tradition. I would add that a tradition that is not aware of itself as a tradition appears to me to be a thing of the past and that globalization which leads to the encounter with many other traditions leaves such 'isolationism' a very unlikely possibility. So engineering will now see itself as one tradition amongst others, some conflicting, some complimentary. Reactivated tradition is what happens when those relationships with other traditions become more explicit and institutionalized. We have to find ways of dealing with those 'others' different from ourselves and that requires greater reflexivity and self-awareness. I would prefer to talk about reconstructed rather than deconstructed traditions simply because that suggests the process of rebuilding that occurs through those interactions with other traditions. What do we learn from each other and how do we now move forward together? Environmental movements have formed an obvious 'other' for science and engineering but have now become absorbed (largely) into a reconstructed tradition. However, this is not the end of the story and further challenges from this source lie ahead. Traditions need to keep changing if they are to serve any useful purpose and need to keep responding to unwelcome and uncomfortable challenges. I am confident that engineering as a profession will rise to these challenges.

References

S. Alkire, *Valuing Freedoms: Sen's Capability Approach and Poverty Reduction*. Oxford, UK: Oxford University Press, 2005.

Z. Bauman, *Liquid Life*. Cambridge, UK: Polity Press, 2006.

U. Beck, *Risk Society: Towards a New Modernity*. London, UK: Sage Publications, 1992.

U. Beck, *World Risk Society*. Cambridge, UK: Polity Press, 2000.

U. Beck, *Cosmopolitan Vision*. Cambridge, UK: Polity Press, 2006.

U. Beck, A. Giddens, and S. Lash, *Reflexive Modernization: Politics, Tradition and Aesthetics in the Modern Social Order*. Cambridge, UK: Polity Press, 1994.

R. Bhaskar, *Meta-Reality: The Philosophy of meta-Reality, Volume 1: A Philosophy for the Present*. London, UK: Sage, 2002.

P. Bourdieu, *Firing Back: Against the Tyranny of the Market 2*. London, UK: Verso, 2003.

M. Castells, *The Rise of the Network Society: The Information Age: Economy, Society and Culture*, Vol. 1. Oxford, UK: Blackwell, 2000.

S. Critchley, *Ethics, Politics, Subjectivity*. London, UK: Verso, 1999.

Engineering and Technology Board, *Engineering UK 2005: A Statistical Guide to Labour Supply and Demand in Engineering and Technology*, London: www.etechb.co.uk

T. C. Fishman, *China Inc: The Relentless Rise of the Next Great Superpower*. London, UK: Simon and Shuster UK Ltd, 2005.

D. Fontana, *Psychology, Religion, and Spirituality*. Oxford, UK: Blackwell, 2003.

M. Hardt and A. Negri, *Empire*. Cambridge, MA: Harvard University Press, 2000.

M. Hardt and A. Negri, *Multitude*. London, UK: Penguin, 2004.

D. Held, *Global Covenant: The Social Democratic Alternative to the Washington Consensus*. Cambridge, UK: Polity Press, 2004.

D. Held and A. McGrew, Eds., *The Global Transformations Reader: An Introduction to the Globalization Debate*. Cambridge, UK: Polity Press, 2003.

J. Keane, *Global Civil Society?* Cambridge, UK: Cambridge University Press, 2003.

R. Layard, *Happiness: Lessons From a New Science*. London, UK: Penguin, 2006.

J. Reader, *Local Theology: Church and Community in Dialogue*. London, UK: SPCK, 1994.

J. Reader, *Blurred Encounters: A Reasoned Practice of Faith*. Cardiff, UK: Aureus Publishing, 2005.

J. Rosenau, *Distant Proximities: Dynamics Beyond Globalization*. Princeton, NJ: Princeton University Press, 2003.

Select Committee on Science and Technology Third Report: United Kingdom Parliament, 23 February 2000. www.publications.parliament.uk/pa/ld199900/ldselect/ldsctech/38/3802

R. Sennett, *The Culture of the New Capitalism*. New Haven, CT: Yale University Press, 2006.

J. Stiglitz, *Globalization and its Discontents*. London, UK: Penguin, 2002.

J. Stiglitz, *The Roaring Nineties: Why We're Paying the Price for the Greediest Decade in History*. London, UK: Penguin, 2003.

United States International Trade Commission, *Tools, Dies and Industrial Moulds: Competitive Conditions in the United States and Selected Foreign Markets*, Investigation No 332-435, USITC Publication 3556, October 2002.

K. Wilber, *Integral Psychology: Consciousness, Spirit, Psychology, Therapy*. Boston, MA: Shambhala Publications, 1999.

Biography

John Reader is a writer and social theorist in the UK. He has degrees from the Universities of Oxford, Manchester and Wales, where he is also an Honorary Research Fellow. He has published a number of books and articles over the last 16 years. As well as working as a parish priest he serves as a board member on a not-for-profit housing organization, regularly visits a business based on scientific research in the global defence industry, and engages far and wide with the impacts of globalization on the lives of the people around him.